D0852836

ADVANCED QFD

ADVANCED QFD

Linking Technology to Market and Company Needs

M. LARRY SHILLITO
Rochester, New York

A Wiley-Interscience Publication
JOHN WILEY & SONS, INC.
New York / Chichester / Brisbane / Toronto / Singapore

This text is printed on acid-free paper.

Library of Congress Cataloging-in-Publication Data:

Shillito, M. Larry, 1939–
Advanced QFD : linking technology to market and company needs / M. Larry Shillito.
p. cm.
ISBN 0-471-03377-4 (alk. paper) :
1. Quality function deployment. I. Title.
TS156.S469 1995
658.5'62—dc20 94-8288

Printed in the United States of America

10 9 8 7 6 5 4 3 2 1

PREFACE

With the total quality management (TQM) movement in the United States there is an ever increasing emphasis on and need for compressing the development and manufacturing cycles. This, in turn, creates a need for tools and techniques to help companies meet the challenge. One particularly useful technique is quality function development, more commonly referred to as QFD.

This book is written for QFD practitioners. It assumes the reader has completed a QFD practitioner course. If one does not use QFD regularly, expertise begins to wane. Likewise, when returning from a cram course in QFD, everything tends to blur together. It is in these situations that a book like this can be useful and appreciated.

I include in this book many hints and recommendations collected from my many successful and unsuccessful applications. With the format used and the subjects covered, this book can also be used as a textbook and training aid.

The first chapter of the book contains a history of QFD. It also explains the basic QFD process by using a four-matrix model that will serve as the basic model for the rest of this book.

Chapter 2 documents in cookbook fashion how to construct the four QFD matrices. The chapter can be used as a checklist to take along to QFD team meetings. Additional operational guidelines are given for each matrix and are located at the end of each matrix section.

Once I have the basics explained, I then take you on a creative excursion, in Chapter 3, into expanding the basic matrices, mainly the house of quality (HOQ), by utilizing other rating schemes and subjects. I also display the basics of value engineering through the use of value graphing to plot the various parameters that may produce a different perspective on column and row elements.

In Chapter 4, I also introduce a hybrid process: customer-oriented product concepting (COPC). This is a function-technology matrix approach for developing products. Customer, market, manufacturing, and design needs are used to optimize new product design concepts. It combines several QFD matrices with value engineering.

In Chapter 5, I integrate QFD with strategic planning. I show how the QFD interconnected matrix chain process can be used in a pre-QFD or pre-HOQ capacity. I have termed the process PQFD for planning QFD. The PQFD matrices will help one to plan for QFD by starting with the corporate mission and working down through three planning matrices to the HOQ. The matrices will help us become better connected to the corporate mission and business plan. They will also help us develop better voice of the customer (VOC) data to start the HOQ.

In Chapter 6, I describe the use of technological forecasting (TF) tools that can be used to measure the effects of signals identified in the PQFD and QFD process. Specifically, I describe the use of the molecular explosion model (MEM) to determine the magnitude and direction of the possible implications resulting from events and actions derived from various QFD and PQFD matrices. Next, I describe the use of an impact matrix to estimate the impact of QFD and PQFD events on the company mission and business plan. Also discussed is monitoring to track signals into the future. Finally, I describe how Delphi inquiry can be used to track VOC and other data into the future. These TF tools help us to recognize and judge the consequences, good and bad, of QFD data.

Chapter 7 describes how to organize and launch a QFD project. The chapter serves as another checklist for practitioners to use in planning and launching a QFD project. A set of 12 time-tested basic starting questions is provided along with a template flow diagram.

Chapter 8 deals with the importance of the behavioral and organizational aspects of QFD. The success of QFD is dependent upon how people, politics, and organizations are considered and integrated into applications.

Chapter 9 discusses the VOC, including when and how to start gathering data, how to structure the data, and how to listen carefully. Tools for classifying and expanding the VOC are considered also.

Chapter 10 is an accumulation of parting subjects and comments about QFD in general.

I have not discussed QFD software in this book. There are several commercial packages available that are quite good. Many companies have developed their own. Many of the electronic spreadsheets that are readily available also lend themsleves quite nicely to building the matrices. All QFD practitioners seem to have their own preferences for software. Consequently, I have purposely avoided the subject.

In writing this book I hope to infuse value thinking into practitioners of QFD. I hope I can take the reader beyond the basics and into the world of planning. I am convinced that QFD should form the basis of prioritization and decision making and can lead to innovative solutions to complex problems. QFD, through linking people across functional areas, can weld a strong link between planning, productivity, and quality.

The examples used in the book are hypothetical and were developed especially for constructing the matrices and graphs in order to better develop understanding of some of the enhancements to the basic QFD process. This book is both paradigm enhancing and paradigm shifting with regard to the models and their applications. In this respect, it is paradigm enhancing by expanding the current matrices with regard to rating variables and plotting schemes as discussed in Chapters 3 and 4. It is paradigm shifting by applying the interlocking matrix technique to planning (Chapter 5), and TF tools to observe the impacts of QFD (Chapter 6).

The book is not meant to be a definitive text on QFD. It is meant to be a seedbed for ideas for expanding QFD beyond its original boundaries and applications.

I hope, through this book, that I can encourage the reader to venture beyond their current QFD paradigms.

So, let us explore the exciting world of QFD!

Rochester, New York
July 1994

M. LARRY SHILLITO

CONTENTS

ADVANCED QFD

CHAPTER 1

QUALITY FUNCTION DEPLOYMENT

HISTORY

Quality function deployment (QFD) originated in Japan and was first introduced at the Kobe shipyards of Mitsubishi Heavy Industries Ltd. around 1972. Its application has since proliferated in Japan, but its use is still not universal. The majority of QFD applications are centered in the high-tech and transportation industries and it is applied primarily in those companies whose products represent a significant part of Japan's export business.[1]

Because quality circles were already well established in companies and because employees were well versed in statistical quality techniques, the groundwork was laid nicely for introducing QFD for developing a competitive advantage in quality, cost, development cycle time, and delivery.

The U.S. exposure to QFD was in 1983 through an article in *Quality Progress* by Kogure and Akao[2] and through Ford Motor Company and the Cambridge Corporation, an international management consulting firm.[3] The interest in and application of QFD in the United States is growing at an incredible rate, despite its brief history.

The two major training sources in QFD are GOAL/QPC in Methuen, Massachusetts, and the American Supplier Institute (ASI) in Dearborn, Michigan. Both sources have developed their own, but similar, QFD models. The ASI uses a basic four-matrix method developed by Macabe, a Japanese reliability engineer. GOAL/QPC advocates a multiple matrix method developed by Akao and incorporates many disciplines in a less structured format consisting of a matrix of matrices. Akao has collected the multiple matrix applications from many Japanese practitioners and assembled them into a new book, *Quality Function Deployment.*[4]

OVERVIEW

Before describing the QFD process in detail it is useful to briefly describe QFD. Quality function deployment is an interdisciplinary team process to plan and design new or improved products or services in a way that:

1. Focuses on customer requirements.
2. Uses competitive environment and marketing potential to prioritize design goals.
3. Uses and strengthens interfunctional teamwork.
4. Provides flexible easy-to-assimilate documentation.
5. Translates soft customer requirements into measurable goals, so that the right products and services are introduced to market faster and correctly the first time.

The QFD methodology consists of a structured multiple matrix-driven process to:

1. Translate customer requirements into engineering or design requirements.
2. Translate engineering or design requirements into product or part characteristics.
3. Translate product or part characteristics into manufacturing operations.
4. Translate manufacturing operations into specific operations and controls.

Based on the ASI model,[5] the translation mechanism is a series of four connected matrices (see Figure 1.1).

The QFD process starts with the needs of the customer and applies them to the entire product life cycle of conceiving, developing, planning, and producing a product or service. Customer requirements are most often referred to as *the VOC*[5] because the requirements are expressed in the customer's own words

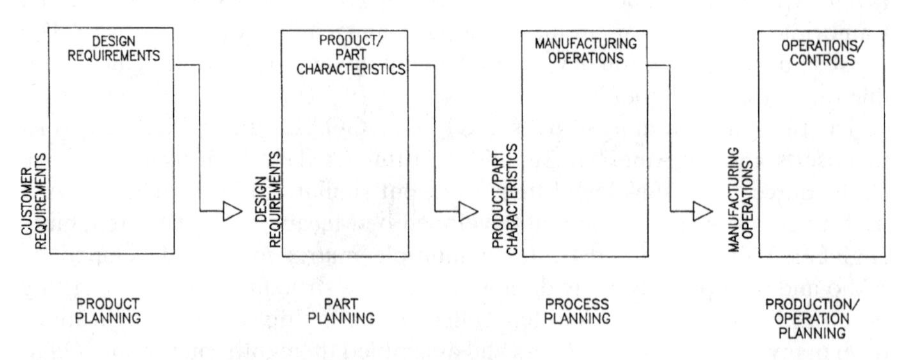

Figure 1.1 The QFD process.

and, in many cases, are written as direct quotes. The process is an attempt to preserve the customer demands by "insulating them against constant reinterpretation throughout the product development cycle and providing a frame work against which future changes can be measured" (Reference 1, p. 2).

PREPARATION

Before describing the construction and use of the QFD matrices, it is necessary to describe three basic structural techniques used to analyze and structure qualitative data. These tools are used to build a matrix of customer information and product features and measures. They are part of a set of seven quality control (QC) tools developed in Japan by the Society of QC Techniques Development, which are known as the *seven new tools.*[6]

The first tool is called an *affinity diagram.* It is used to gather large amounts of qualitative data and organize it into subgroupings based on similarities between items. For example, the QFD team collects and/or generates the needs of the customer for product X. Each individual item (customer need) is written on an index card or sticky notes. All cards are laid at random on a table, or the sticky notes are stuck on a wall or chart board. At this time, the items are unstructured because the data were generated in a random stream-of-consciousness manner. Based on intuition and gut feeling the items are first paired together based on similar attributes and then further aggregated by the team into larger clusters that represent a common theme. Generally items group into 5 to 10 main clusters that each contain 1 to 15 items. The 5 to 10 clusters can be further aggregated many times to form a three- or four-level hierarchy.

To illustrate the affinity diagram we will apply the process to the customer needs of a candle. Based on market surveys, interviews, and/or internal brainstorming, individual customer needs were generated by team members and written on cards (Figure 1.2). The number of items has been considerably reduced for this example. Normally there may be 20 to 80 items. Cards were then rearranged by team members (Figure 1.3). For example, the team decided that "be visually attractive" and "be fragrant" were related to a general theme of aesthetics. In like manner, the other four customers needs were grouped. Note that the team did not develop categories first and then assign the items to the categories. It is more creative and more information-rich to cluster items first and then assign a heading to the cluster. The affinity diagram forces organization, fosters a general level of understanding, and surfaces hidden relationships.

The clusters of items from the affinity diagram can then be arranged horizontally into a second tool called a *tree diagram* (Figure 1.4). That is, the affinity diagram, which is based on intuition and gut feeling, is used to construct a tree diagram based more on logic and analytical skills. The various branching levels of the tree diagram are used to search for gaps and omissions in the

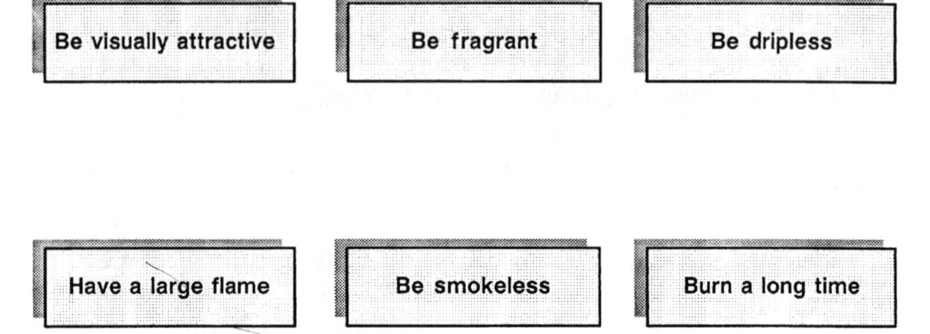

Figure 1.2 Candle user needs (random generation).

affinity diagram. For example, teams often discover new items that were missed during the first generation of needs, as well as uncover new branches and regroupings. The tree diagram allows the team to add, expand, and elaborate on the VOC to form a more complete structure based on interrelationships.

The same procedures are then used to generate product features and measures also called *the voice of the company* (VOC). These measures are also arranged into a tree diagram. The two trees are then arrayed at a right angle to each other, so that a *matrix diagram* can be formed based on the third (or lowest) level items of the two trees (see Figure 1.5). The matrix, the third tool, very

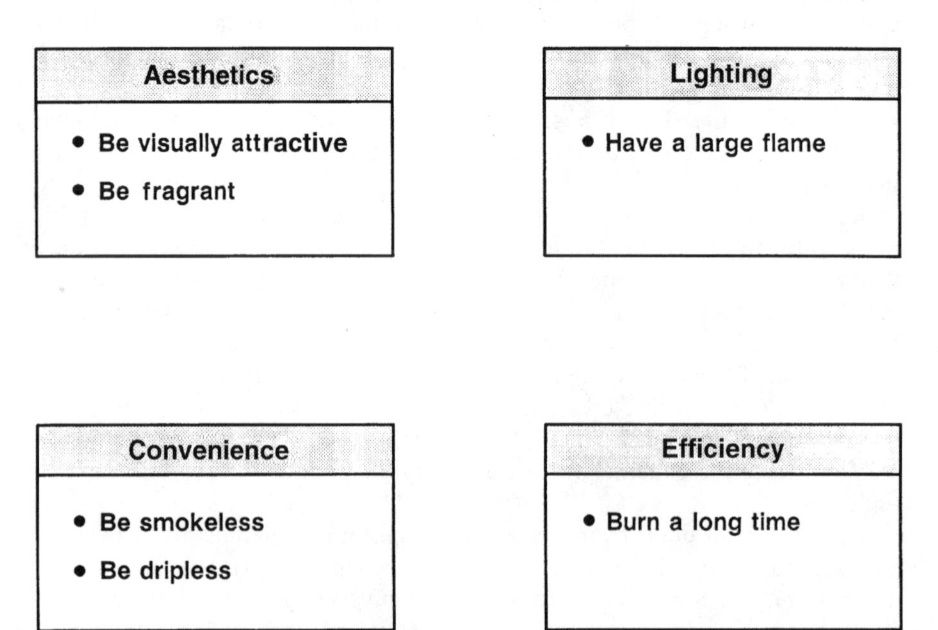

Figure 1.3 Affinity diagram of customer needs.

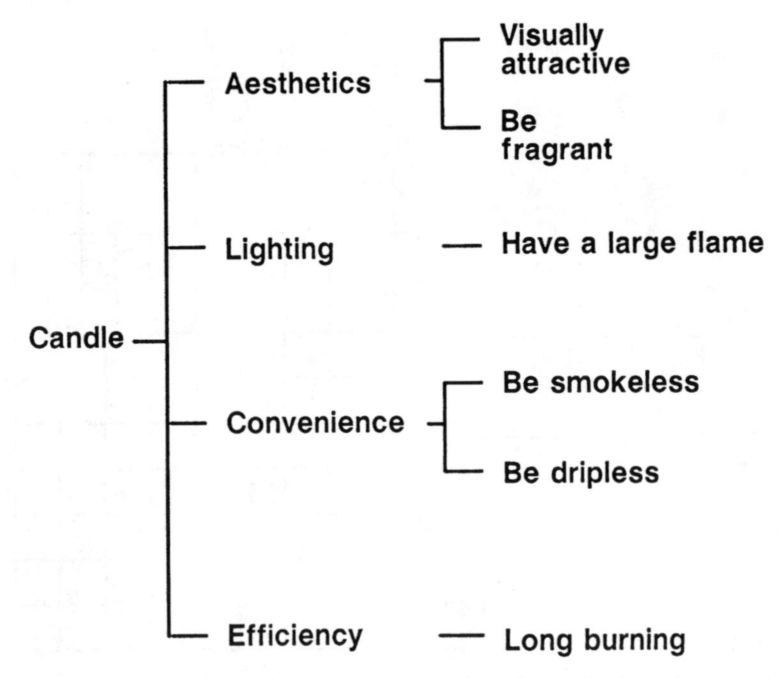

Figure 1.4 Tree diagram of customer needs and technical requirements.

conveniently allows items in one dimension (e.g., voice of the customer) to be mapped against items in another dimension (voice of the company). The matrix provides a structure to systematically evaluate the relationship between the items in both dimensions. Each cell of the matrix provides the basis for asking questions about the relationship between customer needs and product characteristics that may not have been obvious before.

Any relationship between rows and columns may be coded by symbols such as those used by the Japanese. That is, a double circle indicates a strong relationship, usually given a score of 9; a single circle indicates a moderate relationship, given a score of 3; a triangle indicates a possible or low relationship, given a score of 1, and a blank cell represents no relationship.

The intersection of the two trees to form a matrix serves as the basis for constructing the first QFD matrix, termed the house of quality.

THE HOUSE OF QUALITY

The house of quality (HOQ) is the nerve center and the engine that drives the entire QFD process. It is, according to Hauser and Clausing, "a kind of conceptual map that provides the means for interfunctional planning and communications" (Reference 7, p. 63). The HOQ is a large matrix that contains seven different elements (see Figure 1.6).

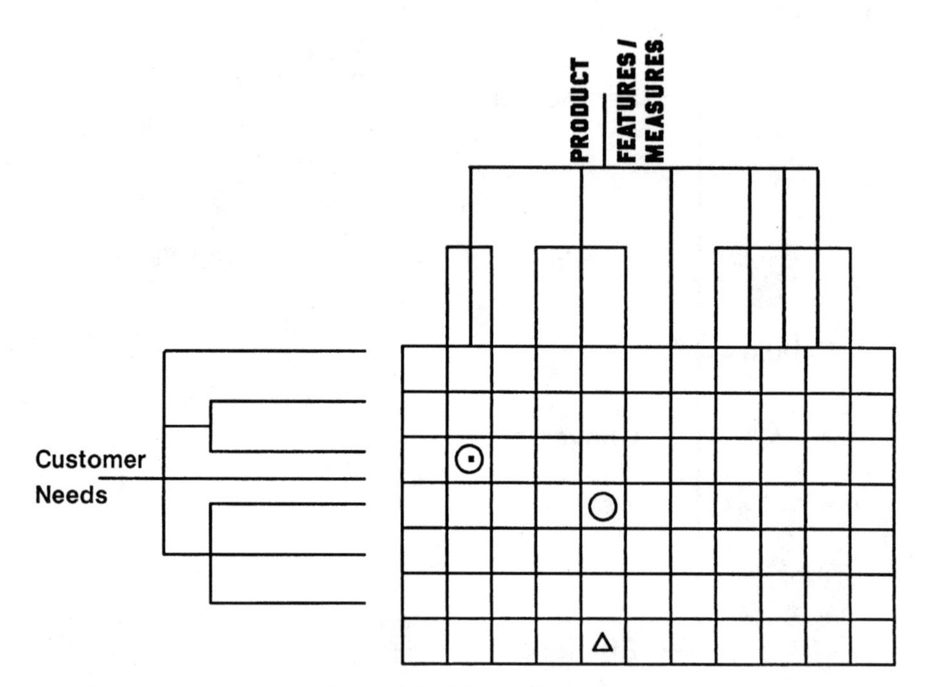

Figure 1.5 Matrix diagram.

The seven elements in the HOQ are the following.

1. *Customer Needs.* These are the VOC. They are also known as *customer attributes, customer requirements,* or *demanded quality.* Very often they are structured by the affinity diagram and the tree diagram. Examples are for a car door, "easy to open"; for a bank, "no waiting in lines"; for a lawnmower, "easy to start." QFD is used to deploy the VOC. It is not used to gather or collect the VOC. VOC is a separate process and is discussed in Chapter 9.

2. *Product Features.* These are also called *design requirements, engineering attributes,* or *substitute quality characteristics.* They, too, can be developed using the affinity diagram and tree diagram. The product features will become the measure to determine how well we satisfy the customer needs. That is, marketing tells us what to do and the engineers/designers tell us how to do it. Examples are for a car door, energy (in ft/lb) to close door; for a bank, computer downtime frequency; for a lawnmower, pulling force to rotate shaft. Product features must be stated in measurable and benchmarkable terms. Many teams use the customer needs to generate the measurable characteristics.

3. *Importance of Customer Needs.* Not only do we need to know what the customer wants, but also how important those needs are.

4. *Planning Matrix.* This portion of the HOQ contains a competitive analysis of the company's product with major competitors' products for each cus-

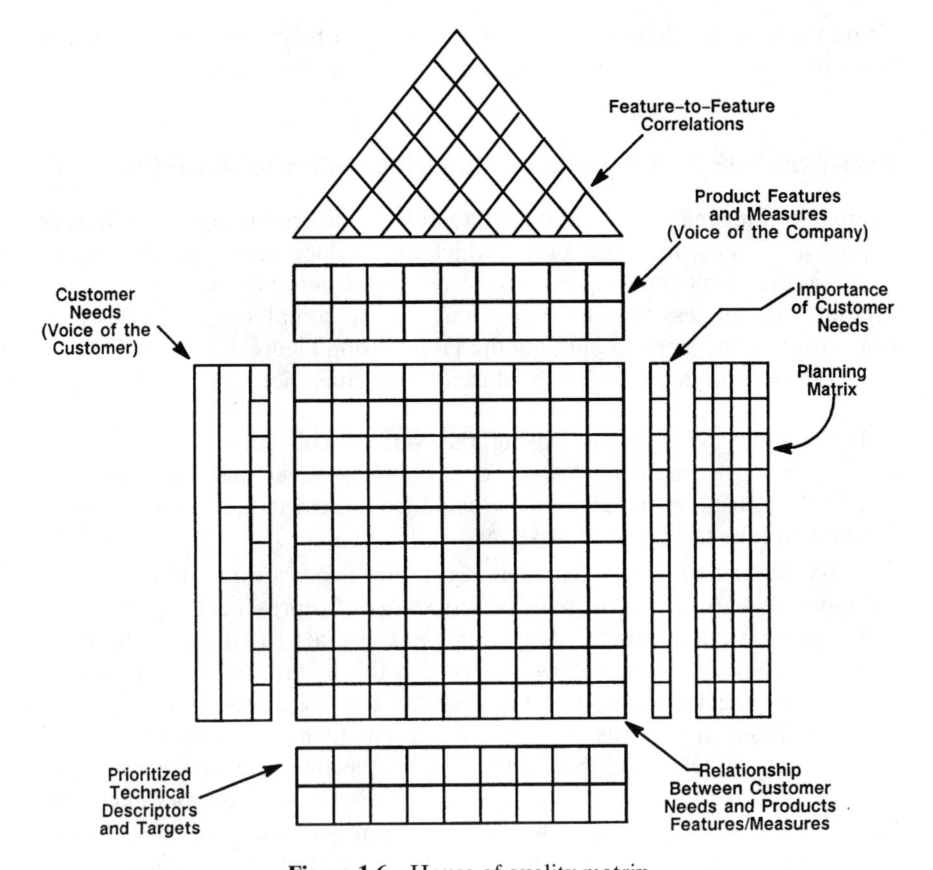

Figure 1.6 House of quality matrix.

tomer need. There are columns to judge how much improvement is needed in the current product, how much sales leverage may result from the improvements, and a final overall score for each customer need. Each score is factored by customer importance, needed improvement, and sales leverage.

5. *Relationship Between Customer Needs and Product Features.* The interfunctional team uses the body of the matrix to indicate how much each product feature (engineering characteristic) affects each customer need.

6. *Feature-to-Feature Correlation.* To what extent will a change in one feature affect other features? Too often a desirable change in one feature results in a negative effect on another feature. This correlation allows one to identify such impacts and tradeoffs.

7. *Prioritized Technical Descriptors or Targets.* This section is a summation of the effects of all prior variables on each product feature. It may also contain target measures for engineering characteristics, as well as a competitive analysis of other manufacturers' measures for the same variables.

Using these seven elements, the HOQ becomes a repository of information that can be used as a mechanism for applying commonsense engineering.

STARTING THE QFD PROCESS—BUILDING THE HOUSE OF QUALITY

There is nothing magical about the HOQ. It is not difficult to construct. It does require a fair amount of team time, which may induce anxiety on the part of the team members and be perceived as a heavy front-end load. As the team gets into the process they become committed to completing it. Let us now work through the steps in building the HOQ, using Figure 1.7 as a reference. The example product will again be the candle used earlier.

1. *Customer Needs/Wants.* Using the various collection methods mentioned previously, the team collects data about what the customers want. A hierarchy of customer needs is constructed from these data using the affinity diagram and the tree diagram described earlier.

2. *Importance of Customer Needs.* Customer needs are rated by the team for importance using a rating scale of 1 to 5 where 5 represents very important and 1 signifies unimportant. Many teams also use a 1 to 10 scale. The basic question being answered is: How important is this feature to the user? Rating the customer needs creates considerable constructive dialogue among team members. Many viewpoints are expressed and many notes are taken. Most of the literature on QFD and the HOQ states that a consensus must be reached among team members. This is great but could consume considerable time. Sometimes a majority decision with a minority report is a good compromise. When rating importance, the team must establish whose point of view the ratings represent; which market, segment, user, or purchaser. Sometimes customers sit on the QFD team to help establish importance.

3. *Competitive Analysis.* The company's current candle product is rated against each customer need, using a 1 to 5 or a 1 to 10 scale. The basic question being answered here is: How well does this manufacturer satisfy the listed feature(s)? That is, how well does our current candle meet the customer's needs. In Figure 1.7 this is represented in the column labeled "Us today." Next, the competitors' candle products are rated using the same scale (columns labeled "OM1" and "OM2").

4. *Future Goal.* Using the same rating scale as above, the team again rates the company candle as to where they desire to be in the future with respect to each customer need (example column "Us in future"). The question addressed when doing so is: How well do we want to satisfy customer needs if we offer the candle we are now studying? In the example, visual attractiveness, with a score of 9, is currently being met with our product and is also in a strong position with other manufacturers' (OM1 and OM2) products. There is no need to change our future candle on this customer need. The need to be dripless also satisfies the customer need but is in a weaker position compared to OM2.

	Visually Attractive	Be Fragrant	Have Large Flame	Be Smokeless	Be Dripless	Long Burning	
Percentage scores	16	15	13	12	8	36	Σ=100
Scores	10	9	8	8	5	23	Σ= 63
Sales Point	1.0	1.2	1.0	1.2	1.0	1.5	
Improvement ratio	1.0	2.0	1.0	1.0	1.1	3.0	
Us in Future	9	8	7	8	9	9	
(other mfg. #2)	8	8	8	7	10	9	
(other mfg. #1)	8	6	7	6	8	9	
Us Today	9	4	7	8	8	3	
IMPORTANCE	10	4	8	6.5	4.6	5	

		SELL FUNCTIONS		USE FUNCTIONS								
		AESTHETICS		LIGHTING	CONVENIENCE		EFFICIENCY					
		Visually Attractive	Be Fragrant	Have Large Flame	Be Smokeless	Be Dripless	Long Burning	Score (sum) Σ = 3451	Percent score	Percent Cost	Measurement Units	Target Units
SCENT	Volatility		⊙					135	3.9	31	Vap Pr.	
	Surface Area	△	⊙					151	4.4		cm^2	
DYE	Concentration	○						48	1.4	9	%	0.2
	Color	⊙						144	4.2		hue	
H^+	Concentration	△		△		⊙	⊙	425	12.3	15	%	10
WICK	Treatment	△		△	⊙		⊙	461	13.4	2	ash wt.	
	Number of Ply	△		⊙	⊙	⊙	⊙	637	18.5		count	5
WAX	Shape	⊙					⊙	468	13.6	43	scale	
	Diameter	△		○		⊙	⊙	451	13.1		cm.	
	Melting Point					○	⊙	348	10.1		Deg.F.	150 F
	Viscosity			○	△	○	○	183	5.3		poise	

Figure 1.7 House of quality matrix for a candle.

Therefore, we would like to be a 9 in the future. There is no need to be a 10 because the team believed that competitor OM2 exceeded the customer need. However, for the need to be long burning, our current product does not satisfy the customer need. Our candle burns too fast and the competitor's candles burn considerably longer. We should make a significant improvement on this customer need. The team decided to be a 9 on this feature so we can be as good as our competitors. They also believed that a 10 would be unnecessary. Likewise, the fragrance of our candle is less than the competitors, so a score of 8 was selected for the future.

5. *Improvement Ratio.* An improvement ratio is computed by dividing the future goal rating (us future) by the current rating (us today). From our example, fragrance has an improvement ratio of 2 and long burning has a 3. These are significant changes from the current candle design.

6. *Sales Point.* The column labeled "Sales point" gives marketing a chance to rate whether or not they would get leverage out of any improvements. That is, would the improvements influence a sale? Sales point is rated as 1.5 indicating a strong or significant sales point, 1.2 denoting a moderate sales point, or 1.0 indicating status quo or no additional sales impact. The basic question being addressed here is: Given the importance of this feature to the customer, and considering the magnitude of the improvement ratio, if we in fact make a change in this feature, can you, marketing, get some leverage from such a change?

7. *Score.* After completing the quantifications in steps 2 through 6, a customer score is computed for each individual customer need by multiplying customer importance, improvement ratio, and sales point. The products of these three numbers provide a hierarchy of customer needs based on the team's assessment of the three variables. These raw scores are finally normalized to percentages (column labeled "% Score"), which will be used later as weighting factors. Although the HOQ is not finished at this point, the information and communication that has resulted so far is more than many companies normally have had. Even stopping here would be time well spent.

8. *Product Features/Engineering Characteristics.* Using team brainstorming and expertise, a list of product features/engineering characteristics (PFEC) is generated. These, too, are arranged into a hierarchy using the affinity diagram and tree diagram. At this point, it is advisable to develop a glossary of terms for each feature. Doing so will make it easier for the team to communicate. Many teams also develop a glossary for customer needs as well.

9. *Customer Needs—Product Features Relationship.* Product features are entered from the tree diagram as columns adjacent to the customer needs. This creates a relationship matrix (Figures 1.6 and 1.7). For each cell of the relationship matrix, the team estimates whether or not there is a relationship between the column and the row. The question addressed is: Will this product feature/engineering characteristic have an effect on satisfying the customer need? The amount of relationship is represented by the following symbols: A

double circle indicates a strong relationship and implies a numerical score of 9; a single circle represents a moderate relationship and uses a score of 3; a triangle signifies a low or possible relationship and is given a score of 1; a blank cell signifies no relationship at all and implies a score of 0. Symbols can be substituted with numbers depending on the preference of the team.

10. *Product Feature/Engineering Characteristic (PFEC) Relationship Score.* In each cell containing a numerical relationship, the relationship score (1, 3, or 9) is multiplied by the corresponding percent score for the corresponding customer need. In our candle example, the cell representing visual attractiveness and shape contains a product of $16 \times 9 = 144$. Likewise, the other relationship under the column shape is multiplied by the other corresponding percent score. A final feature (column) score is calculated by summing all the products in the column. The column totals represent a rank order of PFECs weighted by customer needs. They indicate how much influence the PFECs have on meeting customer needs.

11. *PFEC Correlations.* We are now ready to complete the final part of the HOQ. It is known as the roof because of its triangular shape atop the customer need-product feature relationship matrix (see Figures 1.6 and 1.7). The roof matrix allows the team to identify and quantify the impact, if any, that a change to one PFEC may have on other PFECs. Each cell in one row of the roof matrix represents the intersection of one PFEC with every other PFEC represented in the column headings. The correlations really represent tradeoffs.

As with the relationship matrix in the body of the house, symbols are entered into those cells where a correlation between two PFECs has been identified. Again, a double circle represents a strong correlation, a single circle signifies a moderate correlation, and a triangle denotes a possible correlation. The question being asked when addressing each cell in the roof matrix is: Will a change in the specifications of one PFEC affect the specifications of the other PFEC in the pair? This is equivalent to a cross-impact analysis of PFECs to determine any potential problems that may arise if significant changes are made to any one PFEC. Most teams further designate whether the impact is positive or negative.

Any cell identified with a high correlation is a strong signal to the team, and especially to engineering, that significant communication and coordination are a must if any changes are contemplated in a targeted PFEC. Sometimes an identified change impairs so many others that it is advisable to leave it alone. In our candle, any change in the treatment of the wick will have little effect on any of the other PFECs except for a slight interaction with the number of ply. This is ideal because it is a high scoring PFEC, has little interaction with other PFECs, and relates to four customer needs. It is also a low cost PFEC (discussed later), so any small amount of money spent to change the treatment may benefit four customer needs with few adverse interactions. However, a change in the diameter of the wax can affect six other PFECs. So even though it is a high scoring PFEC that can influence four customer needs, caution should be exercised before any changes are made. Finally, number of ply can significantly affect four customer needs and slightly affect a fifth. It also has a strong interaction with diameter and some interaction with treatment. It too is a likely candidate for change.

The 11 steps complete the construction of the HOQ. However, the literature abounds with many variations. Each company modifies the HOQ process to fit its individual needs. For example, in our candle example, we have added additional information below the PFEC column percent scores. Either real cost or percent cost is added. In our example, cost was aggregated by component.

The current units of measurement for each PFEC may be listed. That is, melting point is measured in degrees Fahrenheit, diameter is measured in centimeters, and so forth. However, how does one measure treatment or type of scent? These PFECs may have to be reworded so that a measure can be used. Remember, earlier we mentioned that PFECs should be measurable. Target values for these measures have also been added. Target measures may also be compared to competitors' products and so forth.

USING THE HOQ

Now that the HOQ is built, what do you do with it? The HOQ is a structured communications device. Obviously it is design oriented and serves as a valuable resource for designers. However, engineers may use it as a way to summarize and convert data into information. Marketing benefits from it because it represents the VOC. Upper management, strategic planners, and marketing or technical intelligence can use it to pinpoint strategic opportunities. Again, the HOQ serves as a vehicle for dialogue to strengthen vertical and horizontal communications. Issues are addressed that may never have surfaced before. The HOQ, through customer needs and competitive analysis, helps to identify the critical technical components that require change. The critical issues will then be driven through the other matrices to identify the critical parts, manufacturing operations, and quality control measures to produce a product that fulfills both customer needs and producer needs within a shorter development cycle time. The matrix can help us buy a look at what it will take to play the game.

OTHER "HOUSES" ON THE BLOCK OF QFD

The same procedures used to build the HOQ may be used to construct other HOQ-type matrices to relate other interconnected variables. The HOQ data can be used to connect other matrices to determine the right parts, the right manufacturing process, the assembly operations, the right quality variables to measure along with the right statistical quality control techniques for measurement. Ultimately, the interconnected matrices lead to the right production plan to build the product.

To do this, the critical columns (PFECs) of the HOQ become the rows of a second matrix that has product/part characteristics as columns. Note that not every column is transferred to the second matrix. Only the issues critical to the success of the product are brought forward. The same procedures used in the HOQ are used to qualify and quantify this new matrix. The PFECs (rows) are used to generate relevant parts characteristics (columns). The purpose of this matrix is to select the right design concepts by determining the critical parts and characteristics. The deliverable emanating from this matrix provides critical data for parts characteristics.

Functional product requirements are often derived by using Function Analysis Systems Technique (FAST) diagrams. FAST diagrams are expanded far enough to produce function characteristics. Product parts can then be assigned to the function blocks of the FAST diagram. These data can then be entered in this second matrix, along with the other design requirements, to isolate critical parts and part characteristics.

This process continues to a third matrix where the columns (critical parts and their characteristics) of the second matrix become the rows of a third matrix. The columns generated for the third matrix contain the key process operations that influence the part characteristics. The objective for using this third matrix is to select the best manufacturing process by determining the critical process operations and parameters. Function analysis and FAST diagrams, again, are excellent tools to determine the manufacturing process steps needing attention. Kaufman[8] is an excellent reference for FAST diagraming. Shillito and DeMarle[9] also provide a good treatise on function analysis and FAST diagraming.

Finally, in the fourth house and last phase of the QFD process, the columns (key process operations) of the third house become the rows of the fourth matrix. The columns of the fourth matrix will contain production operations and controls, such as statistical process control procedures, operator training, mistake proofing, preventive maintenance procedures, and education and training procedures.

The purposes of this last matrix are to determine critical production control and maintenance requirements and the necessary education and training to assure that critical requirements will always be met. The goal is to manufacture a consistent product that, in turn, necessitates minimum production variation. The better a team constructs and completes the three prior houses, the easier it will be to construct the last matrix.

WHAT TO DO WITH ALL THE REAL ESTATE

Because of its breadth, QFD is more than just a quality tool. It is a method for planning that uses many tools to structure interrelationships in order to make decisions about the future. It engenders team work and leadership through focused dialogue.

Engineers and designers are being pushed to think about who their customers are and what those customers need the product to do. This is also a special opportunity for the engineer to become involved. Because QFD starts at the beginning of the product life cycle, there is a greater opportunity for early involvement and a greater opportunity for significant cost prevention.

Although QFD can be a lengthy process, its characteristics appear to be less foreign to participants than those of other quality and value processes. QFD, from the beginning of the HOQ, is a value improvement process that transcends more than cost reduction and quality. The result of a QFD project is a more value-added product for both customer and producer. In this respect, by-products of QFD are reduced lead time, improved quality, reduced costs, and increased market share. These, in turn, affect the bottom line and increase shareholder equity.

QFD will provide the basis for other hybrid value improvement models, such as "customer-oriented product concepting," developed by this author, which is discussed in Chapter 4. I also include it to draw attention to the fact that the practitioner of the QFD process should integrate or adopt other leading-edge value improvement processes to better meet the increasing demand for value-added products and the increasing pressures to develop better value-added manufacturing and production processes.

SUMMARY

In this chapter I have traced the evolution of the application of QFD. For some time there has been a need for customer focused value engineering type methodology. That void is being filled by QFD, which was imported from Japan and is spreading rapidly throughout the United States. QFD complements other quality tools, and those who integrate these tools, both in process and application, will find many rewards in a new world of applications.

REFERENCES

1. Adams, R. M., and Gavoor, M. D., "Quality Function Deployment: Its Promise and Reality," in *Proceedings, 5th GOAL/QPC Conference,* GOAL/QPC, Methuen, Mass., November 1989.
2. Kogure, M., and Akao, Y., "Quality Function Deployment and CWQC in Japan," *Quality Progress,* pp. 25–29, October 1983.
3. Schubert, M. A., "Quality Function Deployment," *Proceedings, Society of American Value Engineers* **24,** 93–98, 1989.
4. Akao, Y. (Ed.), *Quality Function Deployment,* Productivity Press, Cambridge, Mass., 1990.
5. Sullivan, L. P., "Quality Function Deployment," *Quality Progress,* pp. 39–50, 1986.

6. Mizuno, S. (Ed.), *Management for Quality Improvement,* Productivity Press, Cambridge, Mass., 1988.
7. Hauser, J. R., and Clausing, D., "The House of Quality," *Harvard Business Review,* pp. 43–63, May–June 1988.
8. Kaufman, J. J., *Value Engineering for the Practitioner,* North Carolina State University, School of Engineering, Raleigh, N.C., 1985.
9. Shillito, M. L., and DeMarle, D. J., *Value: Its Measurement, Design, and Management,* John Wiley & Sons, New York, 1992.

BIBLIOGRAPHY

1. Bossert, J. L., *Quality Function Deployment: A Practitioner's Approach,* ASQC Quality Press/Marcel Dekker, New York, 1991.
2. Clausing, D., and Pugh, S., "Enhanced Quality Function Deployment," in *Proceedings, Design and Productivity International Conference,* Honolulu, Hawaii, February 1991.
3. Cohen, L. "Quality Function Deployment: An Application Perspective from Digital Equipment Corporation," *National Productivity Review,* Summer 1988.
4. Day, R. G., *Quality Function Deployment,* ASQC Quality Press, Milwaukee, Wisc., 1993.
5. Eureka, W. E., and Ryan, N. E., *The Customer Driven Company: Managerial Perspectives on QFD,* ASI Press, Dearborn, Mich., 1988.
6. Guinta, L., and Praisler, N., *The QFD Book,* AMA Press, New York, 1992.
7. Hertig, J. C., and Abbott, M., "QFD in Pharmaceuticals," in *Proceedings, GOAL/QPC 6th Annual Conference,* Boston, Mass., 1989.
8. King, R., *Better Designs in Half the Time, Implementing QFD Quality Function Deployment in America,* GOAL/QPC, Methuen, Mass., 1987.
9. Marsh, S., Moran, J. W., Nakui, S., and Hoffherr, G., *Facilitating and Training in Quality Function Deployment,* GOAL/QPC, Methuen, Mass., 1991.
10. Pugh, S. "Enhanced Quality Function Deployment in Perspective," in *Proceedings, World Class by Design Conference,* Rochester, N.Y., 1991, pp. 63–77.

CHAPTER 2

CONSTRUCTING THE BASIC QFD MATRICES

BACKGROUND

This chapter outlines in cookbook format the procedures for constructing the four QFD matrices: (1) the product planning matrix, (2) the part planning matrix, (3) the process planning matrix, and (4) the production planning matrix.

Additional operational guidelines are given for each matrix and are located in an operational notes section at the end of each matrix section. It is intended that this chapter serve as a facilitator's guidebook and checklist for conducting QFD projects and leading QFD teams.

Please recognize, however, that this chapter covers the basics. There will be occasions where the basic format will not have an exact fit with the current project and team agenda. The key is to be flexible and creative in the use and application of the process. To help in this matter, I have developed expansions to the matrices as well as modifications to the process, which are discussed in Chapters 3, 4, and 5.

MATRIX 1

House of Quality (Figure 2.1)

I. Purpose:

A. Identify/evaluate customer needs.
B. Identify relationship of customer needs to product technical requirements (PTRs).
C. Evaluate PTRs.
D. Set target values for PTRs.

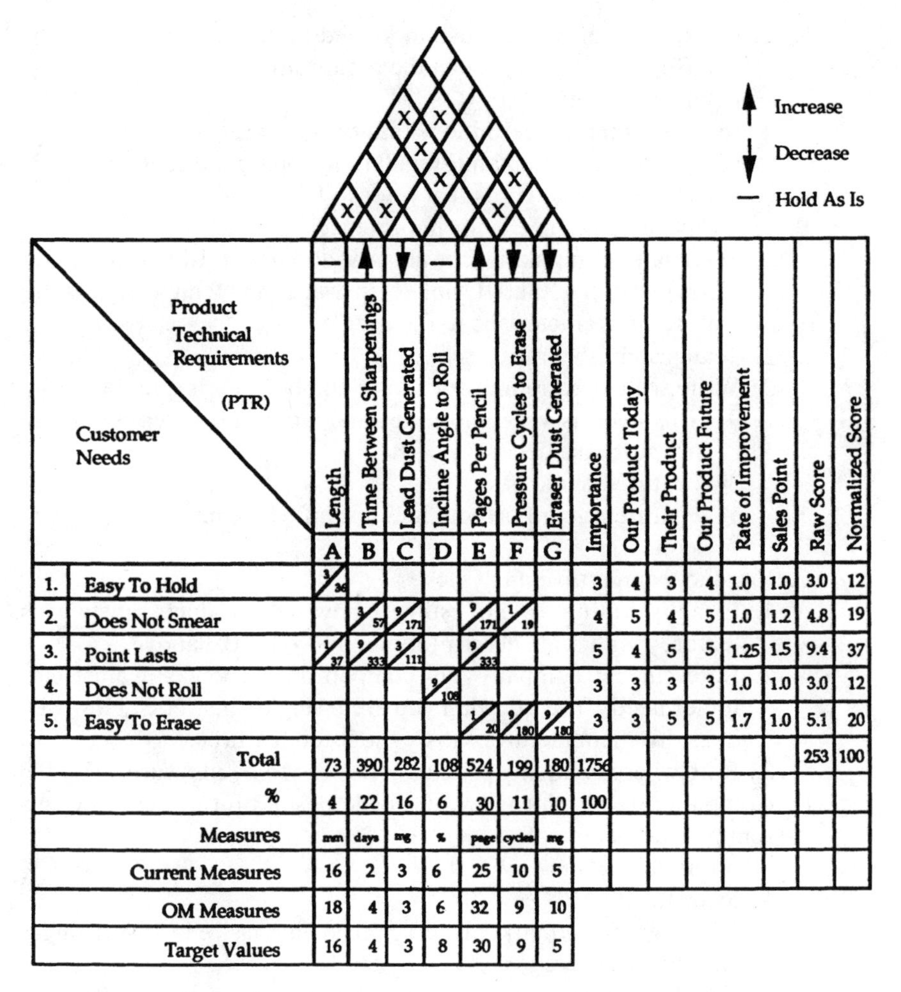

Figure 2.1 Matrix 1, the house of quality for a pencil.

II. Products (Deliverables):

A. Customer needs document.
B. Technical requirements and target measures.
C. Market needs.
D. Product attributes.
E. Competitive analysis.

III. Method:

A. List customer needs:
1. Use whatever sources are available: team, marketing surveys, focus groups, etc. *1,2** (See Chapter 1 for background on the tools used to construct the House of Quality.)

* Numbers refer to notes in Operational Notes section at end of section on House of Quality.

2. Guidelines for writing customer needs: Each statement should:
   (a) Be clear to everyone in the organization.
   (b) Be positive.
   (c) Use common definitions that the QFD team understands.
   (d) Be clearly stated and singular; no compound sentences.
   (e) Be devoid of words that refer to PTRs.

B. Quantify customer needs for importance *3,4:*
   1. Segregate all regulations, warranty, and *musts.* Rate only wants. (The question you are trying to answer is: How important is this feature or characteristic to the user?)
   2. Team establish a rating scale.
   3. Write scale descriptions for rating numbers. For example *5:*
      9 This need is very important and would positively influence my purchase decision.
      3 Feature nice to have.
      1 Feature not important, does not influence my purchase decision.

C. Conduct competitive analysis:
   1. Establish rating scale to estimate how well products satisfy each feature; write scale descriptors similar to III.B.3 above. *6*
   2. Rate current company and competitors' products against customer needs. The question you are trying to answer is: How well does this manufacturer satisfy the listed feature(s)?
   3. Rate where company product is desired to be in future. *7*

D. Compute rate of improvement; divide desired future score by current product score. *8*

E. Quantify sales point *9:*
   1. Sales Used *10:*
      1.5 Significant market leverage (can emphasize in advertising).
      1.2 Some leverage.
      1.0 Status quo.

F. Compute customer need raw score:
   1. Customer need score = Customer importance × Rate of improvement × Sales point.
   2. Normalize raw scores to percentage.

G. List product technical requirements (PTRs) *11:*
   1. PTRs should be measurable and benchmarkable.
   2. PTRs are things we can control in design or manufacturing.
   3. Guidelines for writing PTR statements: GOAL/QPC (Reference 1) recommends that each PTR statement should:
      (a) Be devoid of words that refer to: means, cost, price, reliability, customer demands, parts, tests, process steps.
      (b) Not list parts and processes because we are not trying to design the product at this point.

H. Determine relationship between customer needs and PTRs. *12*
   1. May use symbols: $Q = 9$, $O = 3$, $s = 1$; or words: high = 9, medium = 3, low = 1, definite = 9, some = 3, maybe = 1.

I. Compute PTR scores:
   1. Compute cell scores in relationship matrix:
      (a) Multiply normalized customer need scores (from Step III.F.2) times the cell rating (from Step III.H).
   2. Sum PTR cell scores. This is the column total PTR scores.
   3. Normalize PTR total scores. These percentages represent the relative importance of the PTRs as weighted by the normalized customer importance.

J. Construct the roof of the house (PTR intercorrelations):
   1. For each pair of PTRs, determine whether there is a correlation between them *13:*
      (a) Symbols may be used:
         Q = Strong positive correlation
         O = Positive correlation
         # = Negative correlation
         x = Strong negative correlation

K. Determine measurement units for PTRs and list current measures.

L. Estimate competitors' current values for PTRs.

M. Determine target values for PTRs.

N. Do technical competitive analysis of PTRs. *14*

O. Enter cost, if known, for each PTR.

P. Normalize PTR costs to a percentage.

IV. Tools Appropriate for this Matrix:

A. Affinity diagram.
B. Tree diagram.
C. Matrix diagram.
D. Scaling and scale descriptors.
E. Graphical relationship scale and plot.

V. Issues That Can Be Addressed at this Stage:

A. PTRs.
B. What it takes to enter a market.
C. Market characterization, planning.
D. Selling strategy.
E. Design requirements at macro level.
F. Competitive assessment.
G. Technical benchmarks.
H. See Chapter 3 for optional issues.

Operational Notes:

1. Establish a glossary of terms for all customer needs. As an example, the customer need may be: "sharp image." This may be fur-

ther elaborated so that technical people will understand sharpness. For example, the glossary definition might be: "The image on the paper should have clear definition around the edges. Perception of sharpness may be influenced by contrast." However, the customer's original words still remain intact.

2. An additional structured process for use in defining customer needs is the voice of the customer table (VOCT). The process has just been translated from Japanese and was presented at the QFD Symposium in Novi, Michigan in June, 1991. Two articles relating to VOCT are included in References 1 and 2. The purpose of the VOCT is to provide a medium to explore alternative product application contexts to better define the true needs of the customer as well as explore the unstated "exciting quality" characteristics.
3. Sometimes real customers/users sit on the team.
4. Be sure customer needs importance ratings reflect only one type customer and not a family of customers. If you have multiple customers, create a separate importance rating for each one.
5. Another optional customer importance rating scale used by Ginder[3] is:
   - 5 Critical, would impact the buying decision.
   - 4 Very important, impacts the buying decision when considered with other demands.
   - 3 Like to have, would spend money for it.
   - 2 It would be nice, would like to have, but would not spend money for it.
   - 1 I don't care.
6. An optional company competitive assessment rating scale also used by Ginder[3] is:
   - 5 World class: Meets the expectations beyond the customer's wildest dreams.
   - 4 Best in class: There aren't any better.
   - 3 Average: Pretty good but there is room for improvement.
   - 2 Disappointing: You should have been better.
   - 1 You are in a class by yourself! There is no one as bad as you are.
7. Don't fall into the trap of thinking that you must improve every feature. This will result in overdesign.
8. Sometimes it may be more effective or representative to compute the arithmetic difference between desired future score and current state score. Ratios can sometimes be misleading. For example: an improvement ratio of $3/1 = 3$ is different from an improvement ratio of $9/3 = 3$. In this case, the arithmetic difference is $3 - 1 = 2$ and $9 - 3 = 6$. You will have to be the judge which to use.

9. Marketing personnel must be present to do this rating!
10. Scale values are empirical and are taken from Japanese research.
11. Establish a glossary of terms for all PTRs; this is critical for team communication. For example: "Contrast in toe: Log exposure units between 0.1 and 0.6 density above D-/min."
12. Since matrix 1 can be very large, sometimes only the highest scoring (e.g., top 20%) customer needs are used for the relationship matrix.
13. As in note 12 above, sometimes intercorrelations are done only for the most important 20% of the PTRs.
14. In addition to listing and comparing company and competitor current and target PTR values, many teams use a graphical scale to give a more pictorial representation of the relationships.

MATRIX 2

Parts Deployment (Figure 2.2)

I. Purpose:

A. Determine critical parts.
B. Derive part characteristic values.
C. Identify special attention items.

II. Deliverables:

A. Critical parts and components products document.

III. Method:

A. Enter as rows the most important PTRs from Matrix 1 along with their scores.
B. Add any new PTRs that may also be appropriate at this time. Renormalize all PTR percentages so they add to 100. Note: Rescoring is only done when new PTRs are added.
C. List as columns the parts and components that reflect the new design alternative. These can be generated from a costed bill of materials and/or a FAST diagram. *1*
D. Determine the relationship between PTRs and the parts and components. The same scores (1, 3, 9) are used to indicate the relationship. (See Matrix 1, Step III.H). Enter score on lower half of cell.
E. Compute part/component (PC) scores:
   1. Multiply normalized PTR values (Step III.B) times the parts/component (PC) ratings from Step III.D.
   2. Sum the raw PC cell scores. This is the column total raw PC scores.
   3. Normalize raw PC scores.

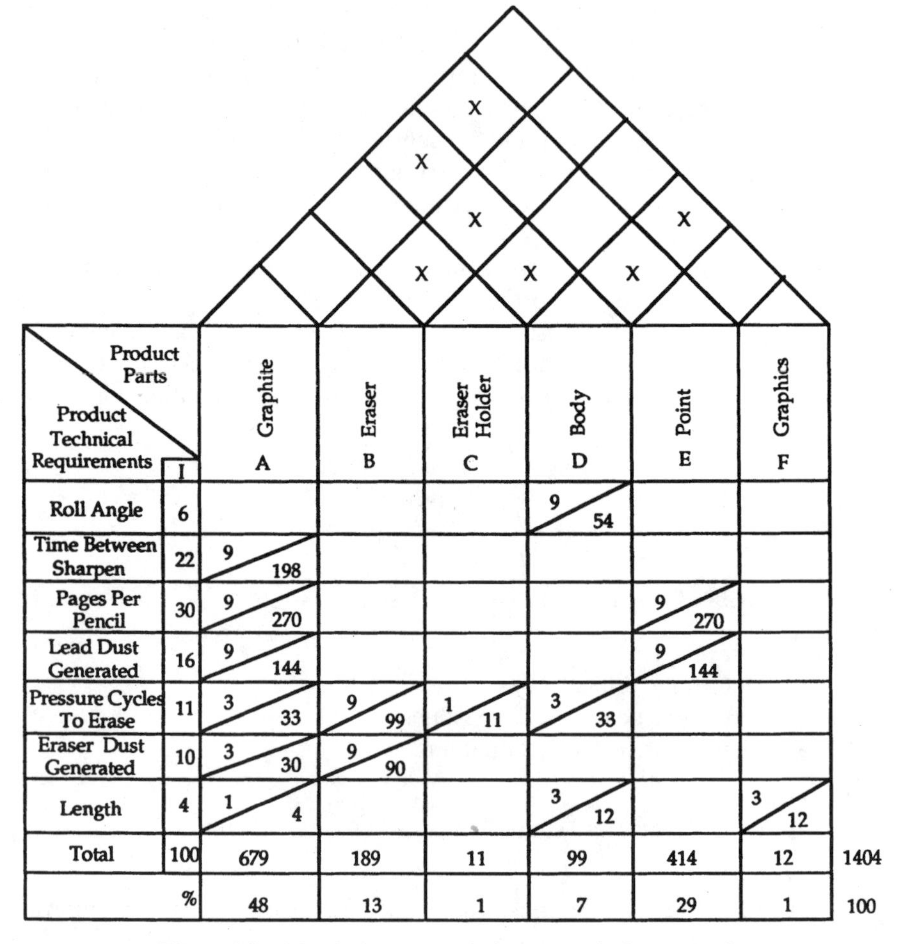

Product Technical Requirements \ Product Parts	I	Graphite A	Eraser B	Eraser Holder C	Body D	Point E	Graphics F	
Roll Angle	6				9 / 54			
Time Between Sharpen	22	9 / 198						
Pages Per Pencil	30	9 / 270				9 / 270		
Lead Dust Generated	16	9 / 144				9 / 144		
Pressure Cycles To Erase	11	3 / 33	9 / 99	1 / 11	3 / 33			
Eraser Dust Generated	10	3 / 30	9 / 90					
Length	4	1 / 4			3 / 12		3 / 12	
Total	100	679	189	11	99	414	12	1404
	%	48	13	1	7	29	1	100

Figure 2.2 Matrix 2, parts and pieces matrix for a pencil.

IV. Tools Appropriate for this Matrix:

A. Matrix diagram.
B. Costed bill of materials.
C. Product explosion diagram.
D. FAST diagram.
E. Pugh concept selection.

V. Issues That Can Be Addressed at this Stage:

A. Material.
B. Hardware.
C. Packaging.
D. Potential suppliers.
E. Subsystems.

F. Product timing.
G. Product tolerances.

VI. Operating Hints:

A. A useful function analysis technique to identify product parts is FAST diagraming as described in Chapter 4, Reference 4, on function analysis.
B. If a design alternative does not exist, it may be necessary to generate one in a creativity session and then use the Pugh concept selection process to narrow the list down to a usable alternative.[5]

Operational Note

1. The PC scores from Matrix 2 are entered as importance factors. Unless new PCs are added to the list, the importance factors do not have to be normalized.

MATRIX 3

Process Planning (Figure 2.3)

I. Purpose:

A. Identify and evaluate process/manufacturing operations (PMOs).
B. Evaluate PMOs for best fit.
C. Determine what things in the process sequence need improvement or redesign in order to make the new concept manufacturable.
D. Determine if processes are adequate.
E. Determine the most important operations.

II. Deliverable:

A. Key process/manufacturing operations.
B. Critical process components.

III. Method:

A. Enter as rows from Matrix 2 the critical part/components (PCs).
B. Also enter the critical PC scores from Matrix 2. *1*
C. Identify process/manufacturing operations needed to produce product. *2* Enter these as new columns.
D. Determine the relationship between critical PCs and the process/manufacturing operations (PMO). The same scores (1, 3, 9) can be used to indicate the relationship (as in Matrices 1 and 2). Enter the score in the lower half of the cell.
E. Compute process/manufacturing operations (PMO) scores:
    1. Multiply the PC scores (Step III.B) times the PMO ratings from Step III.D.
    2. Sum the raw process/manufacturing scores. This is the column total raw process/manufacturing scores.
    3. Normalize raw PMO scores.

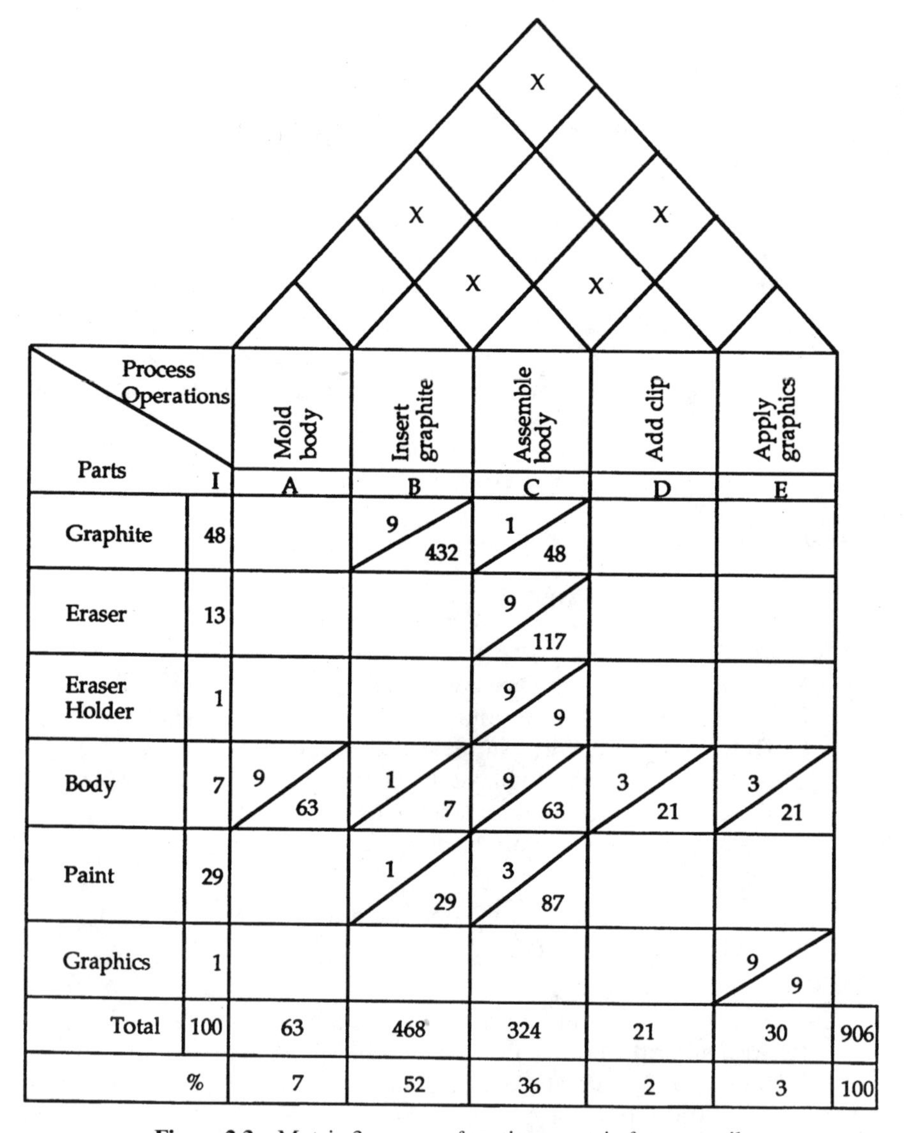

Parts \ Process Operations	I	A: Mold body	B: Insert graphite	C: Assemble body	D: Add clip	E: Apply graphics	
Graphite	48		9 / 432	1 / 48			
Eraser	13			9 / 117			
Eraser Holder	1			9 / 9			
Body	7	9 / 63	1 / 7	9 / 63	3 / 21	3 / 21	
Paint	29		1 / 29	3 / 87			
Graphics	1					9 / 9	
Total	100	63	468	324	21	30	906
%		7	52	36	2	3	100

Figure 2.3 Matrix 3, process functions matrix for a pencil.

F. Review PMO scores to determine what things in the process sequence need improvement or redesign in order to make the new concept parts manufacturable and at the same time satisfy customer needs from Matrix 1.

IV. *Tools Appropriate for this Matrix:*

A. Matrix diagram.
B. Affinity diagram.
C. Production/process operation flow chart.

Issues That Can Be Addressed at this Stage:

A. Process aims.
B. Process tolerances.
C. Process capability index.
D. Tolerance capability index.
E. Process repeatability (in conjunction with Matrix 4).
F. Process reliability (in conjunction with Matrix 4).
G. Critical process components.

Operational Notes:

1. The PC scores from Matrix 2 are entered as importance factors. Unless new PCs are added to the list, the importance factors do not have to be normalized.
2. A process/manufacturing flow diagram should be constructed for each part/component. This exercise will help determine all of the operations, as well as keep the sequence in the proper order. This helps identify what impact each step has to the overall process. The manufacturing or process department should have flow charts of existing operations. These should be helpful in constructing the column titles. If the department flow charts are too detailed, it may be necessary to aggregate operations to a higher level of detail similar to an affinity diagram.

MATRIX 4

Production Planning (Figure 2.4)

I. *Purpose:*

A. Compare process operations to product physical characteristics.
B. Be certain we are checking everything important to a customer.
C. Determine which quality checks are redundant and which checks require more attention.
D. Determine target values for characteristics.
E. Identify internal checks that enable operators to know if the process is operating correctly.

II. *Product/Deliverable:*

A. Key product physical characteristics.
B. Key product characteristics target values.
C. Key internal quality assurance checks.

III. *Method:*

A. Enter as rows the key process operations from Matrix 3. Include also the importance values of those operations. *1*
B. Identify the product physical characteristics. Enter these as new columns. *2* A major part of this can be obtained or derived from Matrix 3.

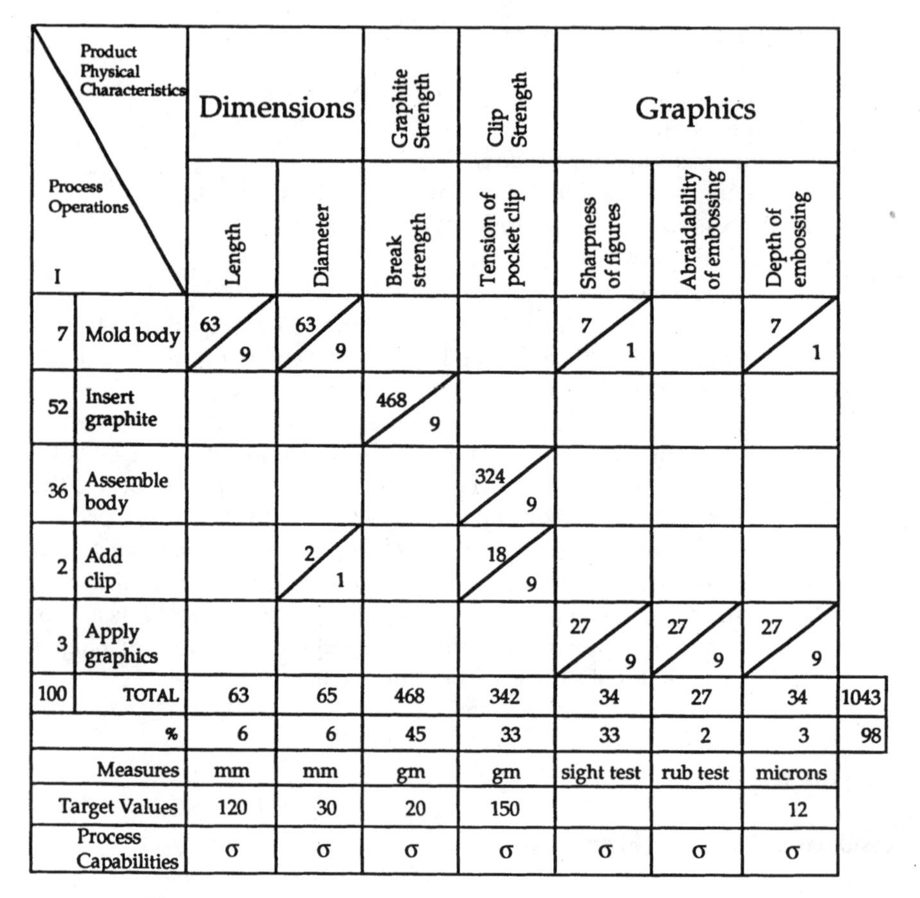

Process Operations I	Product Physical Characteristics	Dimensions		Graphite Strength	Clip Strength	Graphics			
		Length	Diameter	Break strength	Tension of pocket clip	Sharpness of figures	Abraidability of embossing	Depth of embossing	
7	Mold body	63 / 9	63 / 9			7 / 1		7 / 1	
52	Insert graphite			468 / 9					
36	Assemble body				324 / 9				
2	Add clip		2 / 1		18 / 9				
3	Apply graphics					27 / 9	27 / 9	27 / 9	
100	TOTAL	63	65	468	342	34	27	34	1043
	%	6	6	45	33	33	2	3	98
	Measures	mm	mm	gm	gm	sight test	rub test	microns	
	Target Values	120	30	20	150			12	
	Process Capabilities	σ	σ	σ	σ	σ	σ	σ	

Figure 2.4 Matrix 4, production planning matrix for a pencil.

C. Determine the relationship between key process operations and the key product quality characteristics. Score may be 1, 3, or 9 or whatever measure is appropriate. Enter score in lower half of cell. *3*
D. Compute product quality characteristic score, if appropriate. *4*
E. Identify the assembly operations, maintenance requirements, and so on that currently are or need to be addressed as part of quality assurance planning. These are necessary in order to be able to identify the internal checks necessary for operators to control processes.
F. List at the bottom of the matrix all internal checks (test, reading, adjustments) required to maintain process in control. These are part of the quality assurance planning.
G. An intercorrelation "roof" is not constructed because it has already been done in Matrix 1.

IV. Issues That Can Be Addressed at this Stage:

A. Key testing and internal checks for product characteristics.
B. Process repeatability.

C. Process reliability.
D. Critical product and process components.
E. Process aims.
F. Process tolerances.
G. PCIs.
H. TCIs.

Operational Notes:

1. It is likely that not all process operations are carried forward to Matrix 4. Perhaps only the most critical ones are carried forward.
2. You may want to check the product technical requirements (PTRs) and the customer requirements from Matrix 1 to be certain that all of the key product physical characteristics have been identified.
3. Instead of numerical scores, qualifying terms may be used such as high, medium, and low. The purpose is to identify the most critical product quality characteristics needing attention in the manufacturing process.
4. A factored column score may or may not be computed.

SUMMARY

This chapter is to serve as a field manual for QFD practitioners. I have written in cookbook style the step-by-step procedures for constructing the four matrices. Appended to each matrix writeup is a set of operating notes where I have included suggestions, do's and don'ts, and additional information.

REFERENCES

1. Nakui, S., "Comprehensive QFD Systems," in *Proceedings, 3rd Symposium on Quality Function Deployment,* Novi, Mich., June 1991, pp. 132–152.
2. Mazur, G., "Voice of the Customer Analysis and Other Recent QFD Technology," in *Proceedings, 3rd Symposium on Quality Function Deployment,* Novi, Mich., June 1991, pp. 285–289.
3. Ginder, D. A., "The Strategic Approach to Market Research," in *Proceedings, 3rd Symposium on Quality Function Deployment,* Novi, Mich., June 1991, pp. 418–428.
4. Shillito, M. L., and DeMarle, D. J., *Value: Its Measurement, Design, and Management,* John Wiley & Sons, New York, 1992.
5. Pugh, S., *Total Design,* Addison-Wesley, Reading, Mass., 1990.

CHAPTER 3

CREATIVE EXPANSION OF THE HOUSE OF QUALITY AND THE OTHER MATRICES

THE "BASEMENT" (COLUMNS) OF THE HOQ

In Chapter 2 we constructed the basic house of quality (HOQ). The purpose of building the HOQ is to determine the most important customer needs, which in turn allows us to develop a hierarchy of importance for the product technical requirements (PTRs). The PTRs are used to drive forward to Matrices 2, 3, and 4. Various metrics such as other manufacture's current values and your own company's current and target values have traditionally been used in order to isolate the vital few to be carried forward. The literature and courses abound with other measurements and subjective estimates that can be applied at the lower part of the HOQ. This lower portion is commonly referred to as the "basement."

The purpose for expanding the matrices is to use creativity to encourage the QFD team to go beyond the basics. It is the use of other metrics in an information searching mode to discover relationships and interactions.

Let us observe some other metrics that can be applied to the column PTRs in the basement of the HOQ. Refer to Figure 3.1 for this discussion.

1. *Cost.* When known, real cost can be allocated to the PTRs (known as hard cost). If cost is not known, it can be subjectively estimated using various value measurement techniques.[1] I refer to subjective costs as soft costs. The most preferred subjective rating process is scaled pair comparison. All costs are normalized to a percentage.

2. *Value Index.*[1] The formula for value is

$$V = \%I \div \%C$$

where: %I = relative percent importance
%C = relative percent cost

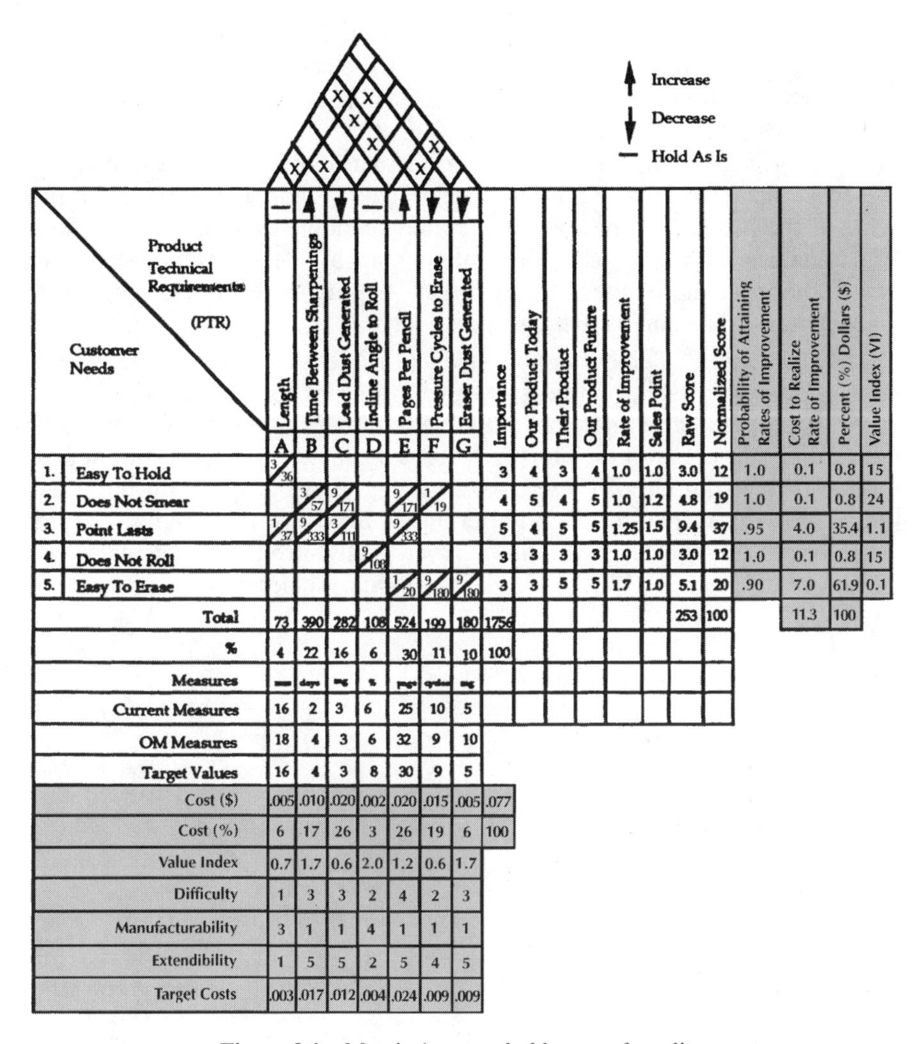

Figure 3.1 Matrix 1, extended house of quality.

To compute a value index for the PTRs, simply divide the PTR relative percent importance by its corresponding relative percent cost. The normalized column totals (Figure 3.1) are used as a surrogate for importance. Real cost was posted to the respective PTR columns and normalized. If hard cost is known, use it; otherwise, subjectively estimate cost and post it along with the respective normalized cost. A value index greater than 1 represents good value, whereas a value index less than 1 represents areas for value improvement. A value index equal to 1 represents break even.

3. *Target Cost.* The value index for each PTR (in 2 above) can be multiplied times the respective actual current cost (in 1 above) to obtain a target cost for each PTR. A target cost less than the actual cost represents areas for value improvement and cost reduction. Sometimes target cost can be higher than the original cost. This would indicate that additional money spent on these PTRs may improve the importance of the PTR and, in turn, increase the value of the product to the user. Sometimes spending more money on these PTRs enables a QFD team to eliminate or improve other PTRs. Be careful before allocating more money. These target costs are merely to challenge your imagination and reasoning. Do they make sense?

4. *Value Graph.*[1] Percent importance and percent cost can also be plotted on a value graph. For example, the value graph for the seven PTRs in our HOQ is represented in Figure 3.2. One's attention is immediately directed to items A, C, and F. Can these plots be a value mismatch? Are costs out of proportion to their importance? Are we spending enough money on the other four items with respect to their importance? Both the value index and the value graph give the team yet another perspective from which to view the PTRs. For example, a PTR that scores low in importance and has a very high

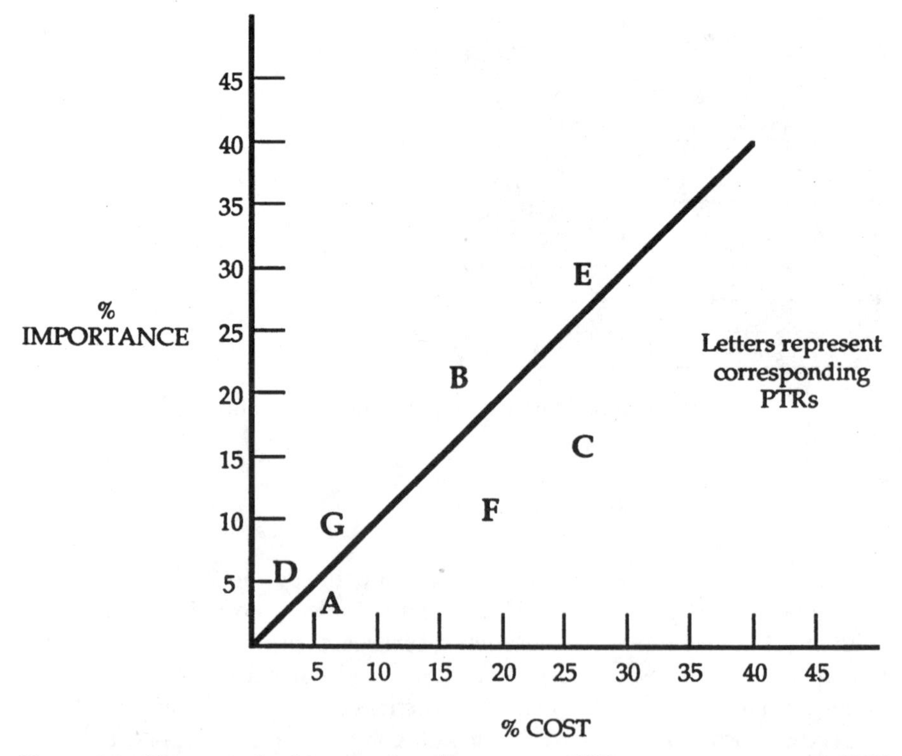

Figure 3.2 Value graph of house of quality, percent PTR score versus percent PTR cost (letters represent corresponding PTR column items).

cost is a candidate for further study. It helps the team prioritize its time. The team should spend the time to rationalize all the plots on the value graph. Such a parity check is a very information rich exercise for the team. New things are uncovered and new perspectives are developed. The basic question is: Do all of the plots make sense? If not, why not? What did we forget or overlook? Do the ratings seem plausible?

5. *Difficulty.* Many teams also rate PTRs for difficulty of changing or maintaining current or target values. A possible difficulty scale might be as follows:

DIFFICULTY RATING SCALE

Score	Descriptor
5	Major difficulty; invention required; unknown technology.
4	Difficult; technology not immediately available; major development work required.
3	Some effort needed; no invention required; some development work needed.
2	Little effort; technology known; some development needed but not major.
1	Easy; known technology; off-the-shelf; plug-in.

This metric, along with all the others, gives the team yet another perspective to carry PTRs forward to the next matrix.

6. *Manufacturability.* Some teams prefer to rate the PTRs for manufacturability. This, of course, depends on the nature of the PTRs. In many cases, rating manufacturability may be premature. It may also be redundant if difficulty has already been rated. Don't forget, manufacturability is dealt with in Matrices 3 and 4. The manufacturability rating, if done in Matrix 1, is done at a macro level to see if some early signals start to surface. A manufacturability rating scale might be as follows:

MANUFACTURABILITY SCALE

Score	Descriptor
5	Difficult; breakthrough needed.
4	
3	Some effort needed; time and hard work will see it through.
2	
1	Easy; no changes needed: can plug into existing lines.

7. *Extendibility.* Extendibility means to what degree can changes in the PTR be extended to other models, product lines, or core technology. For example, a PTR with a high cost but with high importance and extendibility may

more than justify the cost of pursuing it. Extendibility can also be rated, for example:

EXTENDIBILITY RATING SCALE

Score	Descriptor
5	High: good transfer to other products.
4	
3	Some transfer to other lines.
2	
1	No transfer; status quo.

Extendibility starts to enter the twilight zone of core technology, platform centers, and core competency.[2] Using these kinds of measures is applying QFD with a future dimension. It encourages the QFD team to stretch its vision and imagination.

8. *Creative Graphing.* Figure 3.3 represents several plots that can be made using the data from items 1 through 7. Certainly not all of the ratings nor all of the plots would ever be used at the same time. The contents of this chapter are meant to illustrate creative extensions to the basic HOQ. Don't be mentally locked-in to the basic matrix as learned from a course.

THE "MAIN FLOOR" OF THE HOQ

Let us use our imagination and apply similar ratings and operations to the "main floor" (rows) of the house to see if we can derive some early macro level signals for any VOC line items.

1. *Probability.* Probability can be applied not only to the columns, but to the rows of customer needs in the HOQ. Specifically, what many teams do is estimate the likelihood of attaining the calculated rate of improvement (improvement ratio). This factor should be used for discussion only. Some teams do use it as a multiplying factor in calculating the raw customer need score. The reader will have to decide whether or not to use it in this way as yet another, a fourth multiplying factor to compute the raw customer need score. I prefer to let it stand as is and use it as a discussion tool.

2. *Cost.* Another comparison measure, along with probability, would be the estimated cost of achieving the improvement goal for each VOC line item. Now you have at a macro level the guesstimated cost of achieving the VOC goals along with the likelihood of achieving them. Together these give you a preview of what it might take to enter or expand the market by meeting or improving VOC items. This cost is also normalized.

3. *Value Graph.* Once cost is normalized it can be plotted with probability in another value graph (Figure 3.4). In Figure 3.4, VOC item 5 certainly begs

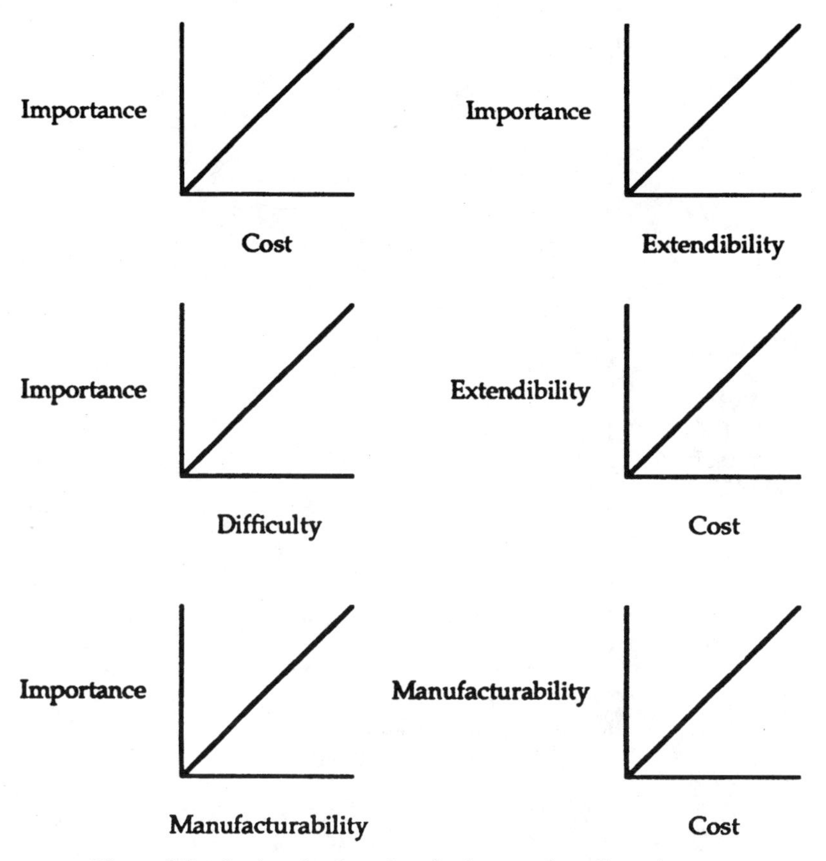

Figure 3.3 Optional value plots for house of quality columns.

for more attention. It should be compared to the customer importance. Likewise, item 3 should be questioned, especially if they have a high improvement ratio. As with the basement of the HOQ, a value index can be computed (see Figure 3.1 rows) for each VOC line item based on the normalized score (%) and the percent cost to realize the rate of improvement. These same data may also be plotted on a value graph as in Figure 3.5. The plot of VOC item number 5 vividly stands out as requiring attention. It is second in importance but is consuming 60% of the cost. How can this be? Does it make sense? How can we get the cost down? What happened? Where do we go from here?

IMPORTANCE VERSUS SATISFACTION

Both of these metrics are already contained in the HOQ under the columns, "importance" and "our product today." Both metrics have already been rated

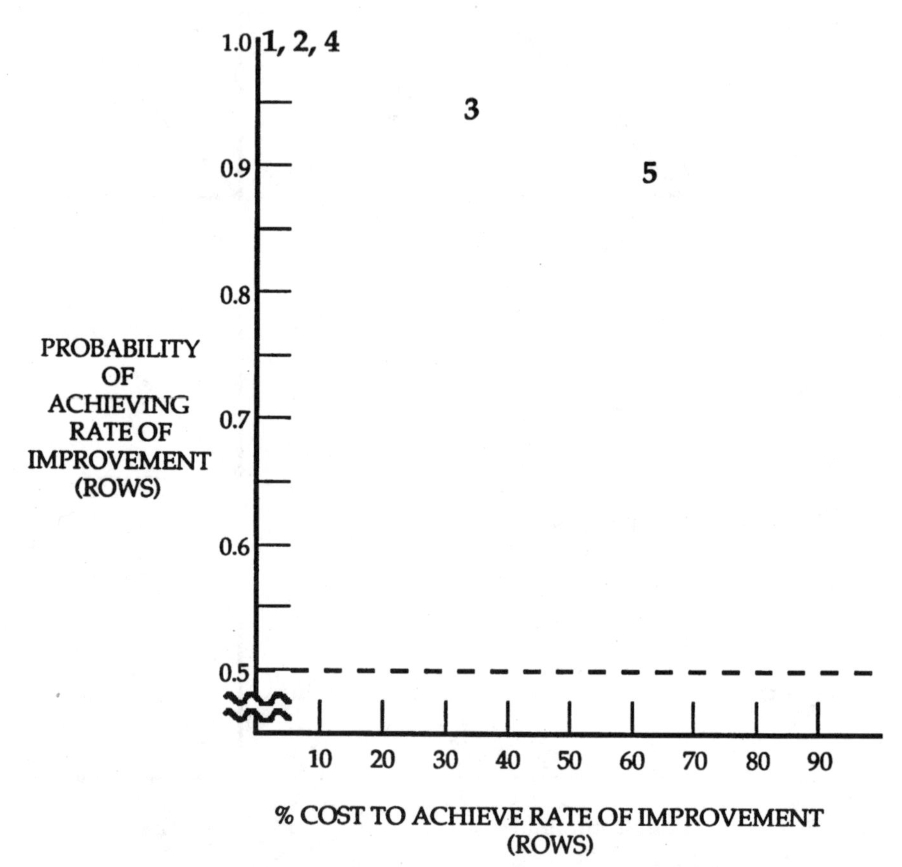

Figure 3.4 Probability-cost plot, probability of achieving rate of improvement for voice of the customer row items versus percent cost to achieve the rate of the improvement. (Numbers represent corresponding VOC line items.)

on a 1 to 5 scale. A graph could be used to display these data, that is, the customer verbatim importance could be plotted against how well our current product is satisfying the customer need (see Figure 3.6). Again, numerical data displayed in graph form can often be more revealing than columns of numbers. Major competitors' satisfaction ratings can also be plotted on the same graph, using a color code to distinguish between positions.

In our graph in Figure 3.6, all of the verbatims scored a "3" or better in importance as well as in satisfaction. Aren't we so lucky! What do you do if some of the items score high in importance and low in satisfaction?

IMPROVING OTHER HOUSES ON THE BLOCK

Likewise, other ratings, ratios, and plots similar to those discussed above, may be used in Matrix 2 (parts and pieces) and Matrix 3 (process functions). For

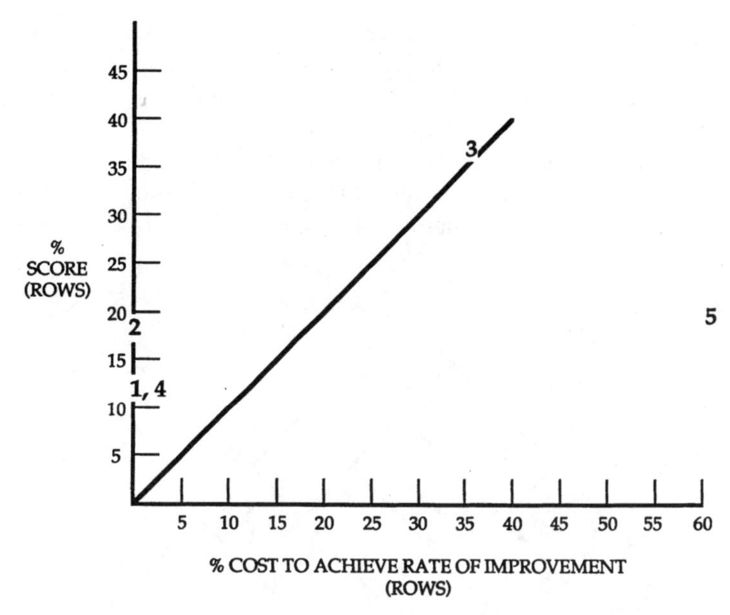

Figure 3.5 Value graph, percent importance of customer needs versus percent cost of achieving rate of improvement.

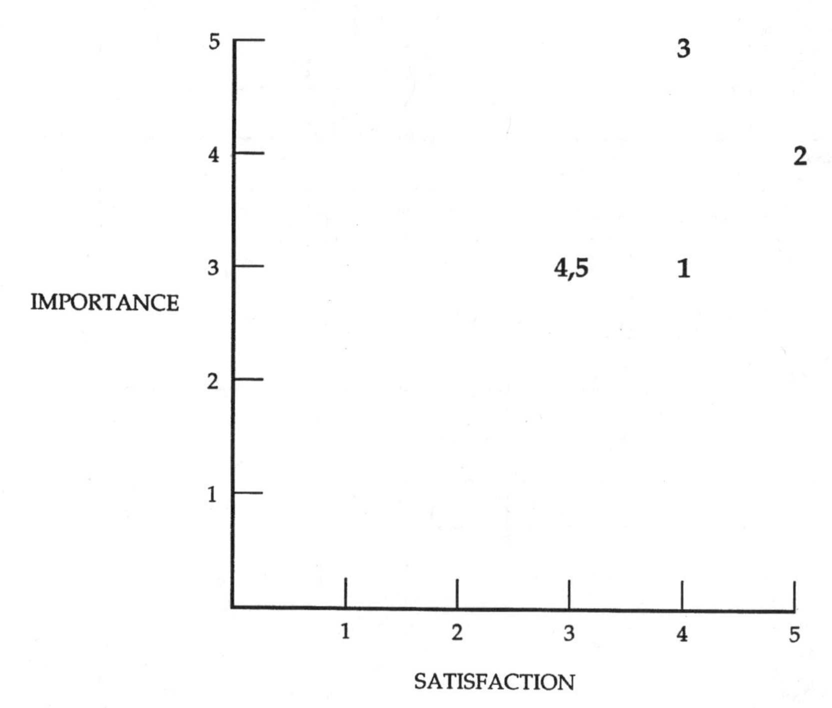

Figure 3.6 Value graph, customer importance versus product satisfaction for VOC verbatims.

example, on Matrix 2 (Figure 3.7) cost and reliability may be estimated for product parts. A value graph can be plotted for part importance and cost, reliability can be plotted against cost, and it can also be plotted against importance. Figure 3.8 illustrates some possible plots for Matrix 2.

Let's extend our imagination to Matrix 3 (Figure 3.9). Here one can rate manufacturability and estimate cost. The importance of process functions can be plotted against their cost. Likewise, manufacturability can be plotted against

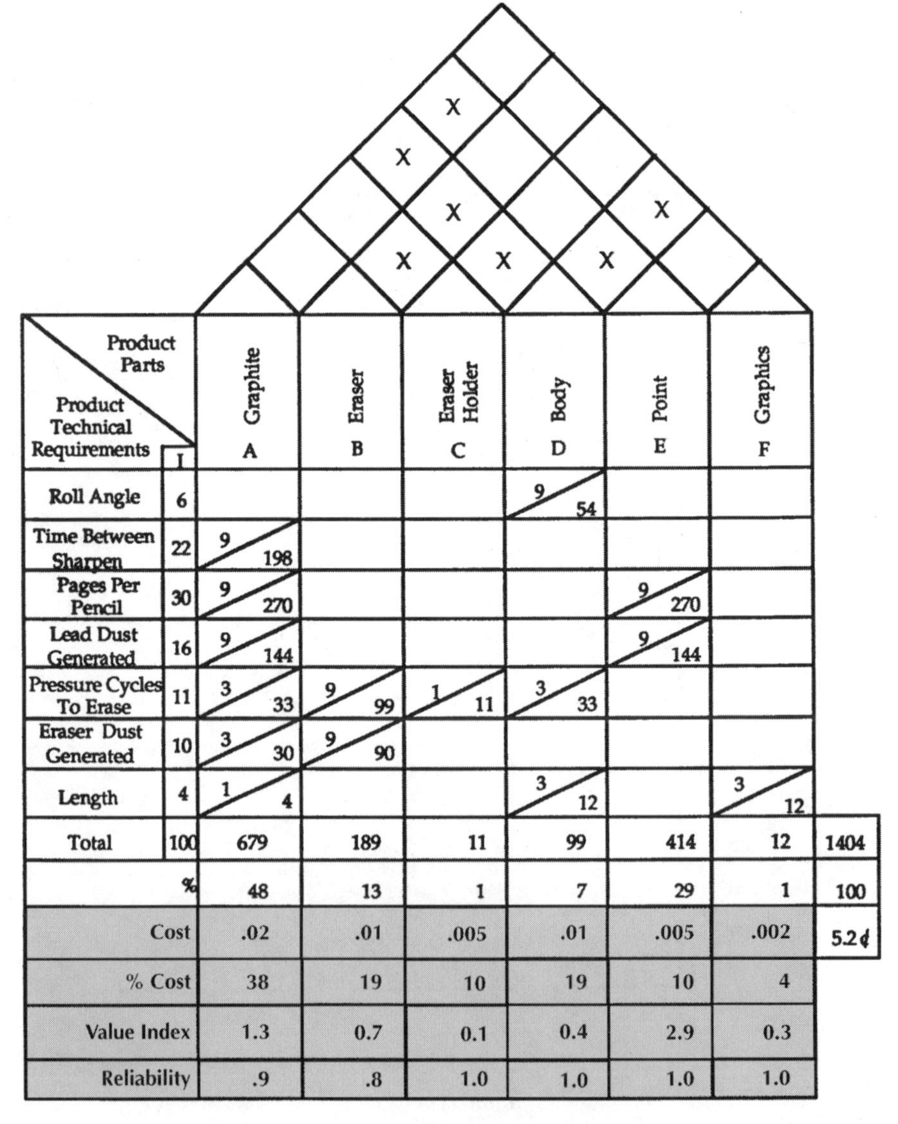

Product Technical Requirements / Product Parts	I	Graphite A	Eraser B	Eraser Holder C	Body D	Point E	Graphics F	
Roll Angle	6				9 / 54			
Time Between Sharpen	22	9 / 198						
Pages Per Pencil	30	9 / 270				9 / 270		
Lead Dust Generated	16	9 / 144				9 / 144		
Pressure Cycles To Erase	11	3 / 33	9 / 99	1 / 11	3 / 33			
Eraser Dust Generated	10	3 / 30	9 / 90					
Length	4	1 / 4			3 / 12		3 / 12	
Total	100	679	189	11	99	414	12	1404
%		48	13	1	7	29	1	100
Cost		.02	.01	.005	.01	.005	.002	5.2¢
% Cost		38	19	10	19	10	4	
Value Index		1.3	0.7	0.1	0.4	2.9	0.3	
Reliability		.9	.8	1.0	1.0	1.0	1.0	

Figure 3.7 Matrix 2, extended parts/pieces matrix.

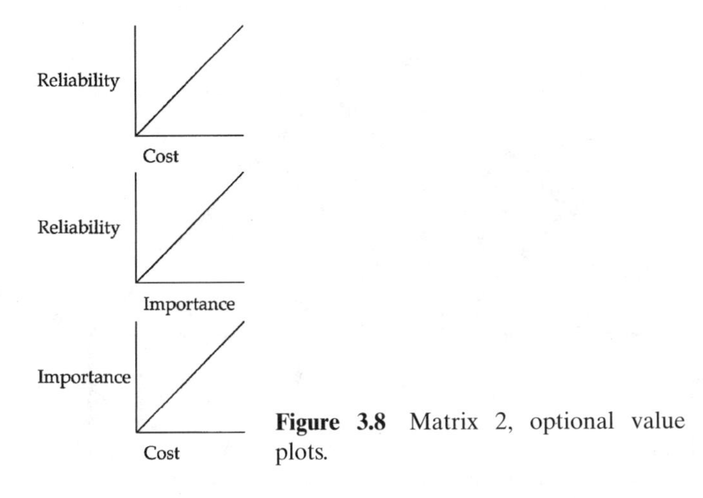

Figure 3.8 Matrix 2, optional value plots.

cost, and process function importance can be plotted against manufacturability. The manufacturability rated in Matrix 3 is much more specific than the macro manufacturability rating in the HOQ. The two sets of ratings should be compared for consistency. Figure 3.10 generically illustrates some of these plots.

THE VALUE GRAPH—SOME COMMENTS

Value graphing was developed by David DeMarle.[1] It is used to plot the relative importance and relative cost of items. Once plotted the graph can be used to locate areas for cost reduction and importance improvement.

Figure 3.11 is a hypothetical value graph used to illustrate value graphing and value targeting. With regard to plot number one, point "A" represents the target cost. It is the intersection of a line parallel to the cost (x) axis and the 45 degree line. Point "B" represents the importance target. It is the intersection of a line parallel to the importance (y) axis and the 45 degree line. Point "C" is the value target and is the perpendicular drawn from the plot origin to the 45 degree line.

To approach "A" requires the use of value engineering and traditional cost reduction. To reach "B" involves adding more functionality and utility while holding cost constant. This can be done by applying VOC, QFD, and COPC (discussed in Chapter 4). The value target is the best compromise between "A" and "B." It is achieved through the creative combination of QFD and COPC in conjunction with VOC, value engineering, and Pugh Concept Selection.

The value graph brings focus to the QFD project and helps to locate areas needing attention and improvement. The use of VOC and QFD help approach these targets.

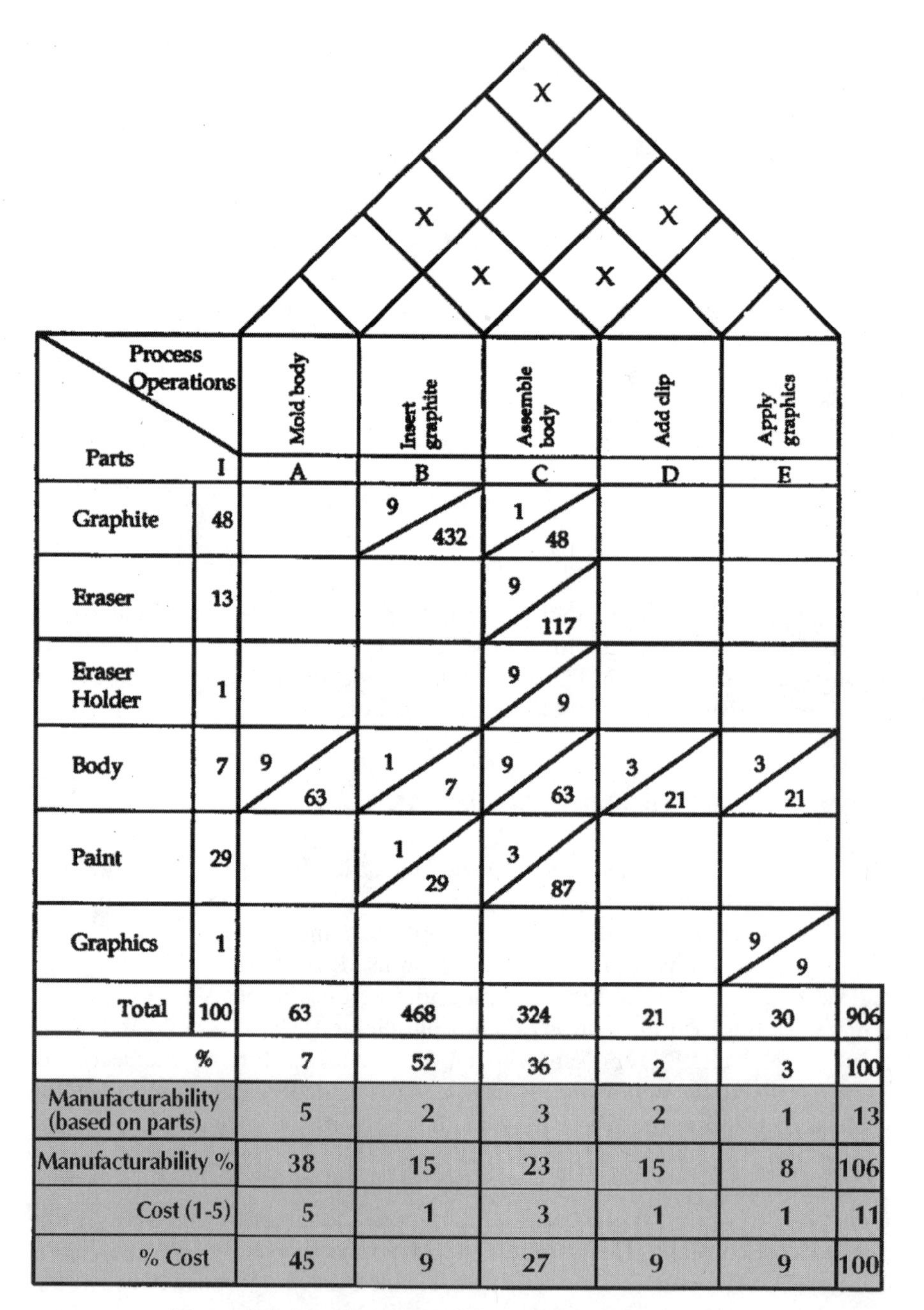

Parts \ Process Operations	I	Mold body (A)	Insert graphite (B)	Assemble body (C)	Add clip (D)	Apply graphics (E)	
Graphite	48		9 / 432	1 / 48			
Eraser	13			9 / 117			
Eraser Holder	1			9 / 9			
Body	7	9 / 63	1 / 7	9 / 63	3 / 21	3 / 21	
Paint	29		1 / 29	3 / 87			
Graphics	1					9 / 9	
Total	100	63	468	324	21	30	906
%		7	52	36	2	3	100
Manufacturability (based on parts)		5	2	3	2	1	13
Manufacturability %		38	15	23	15	8	106
Cost (1-5)		5	1	3	1	1	11
% Cost		45	9	27	9	9	100

Figure 3.9 Matrix 3, extended process functions matrix.

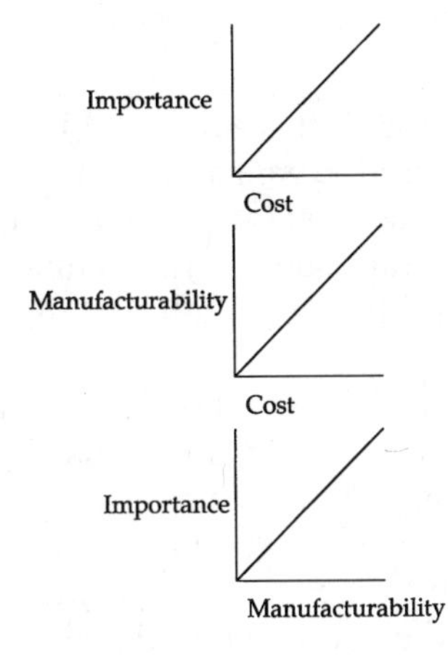

Figure 3.10 Matrix 3, optional value plots.

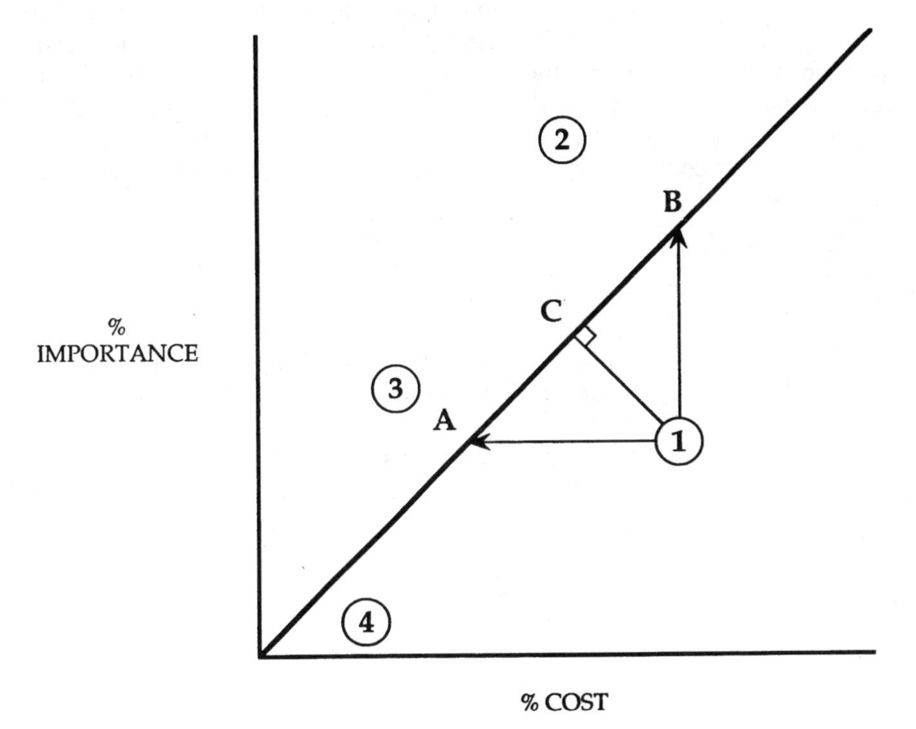

Figure 3.11 Hypothetical Value Graph; A equals the cost target, B is the importance target, and C is the value target.

SUMMARY

So, what do we have? What can we do with all this additional information? The various ratings and graphs can all be used to gain a different perspective on the matrix column and row items. If all the signals point in the same direction or if differences can be explained or rationalized, this should raise the confidence level for proceeding forward in the QFD study and the design of the product.

The additional ratings and measures described in this chapter are generally applied at a macro level of detail, say, level 2 in a tree diagram. Other teams have applied similar measures to only the most important columns or rows, for example, the top Pareto 20%. It is not practical to apply measures to every column or every row because doing so leads to information overload and defeats the purpose of extracting the most important elements in a matrix.

The intent of this chapter is to illustrate the need to go creatively beyond the basics. Not everything in the chapter may be applicable to your particular projects. If not, then so be it. However, I am sure that you, the reader, can come up with your own creative measures and plots that can extend your thinking beyond your usual thinking paradigm. An important point to remember is, don't force the techniques or the measures to fit your project. This will lead to confusion in application and be perceived as technique for technique's sake.

REFERENCES

1. Shillito, M. L., and DeMarle, D. J., *Value: Its Measurement, Design, and Management,* John Wiley & Sons, New York, 1992, Chapter 4 on Value Measurement Techniques.
2. Prahalad, E. K., and Hamel, G., "The Core Competences of the Organization," *Harvard Business Review,* pp. 79–91, May–June 1990.

CHAPTER 4

VARIATIONS ON A THEME: CUSTOMER-ORIENTED PRODUCT CONCEPTING

BACKGROUND

Producing a product "on the first try" that has high end-user value should be a goal of any manufacturer. Such a desire is universal but, unfortunately, not often achieved. Product design is too often executive or producer driven as opposed to customer driven. Such an approach usually leads to massive redesign and/or post-introduction problem solving. Often lacking is sufficient homework in the conceptualization or preliminary design stage in the front end of the product life cycle. This lack of homework merely increases product cost and delivery time.

DEFINITION

A method for assisting design concepting is customer-oriented product concepting (COPC), which is an interdisciplinary team process to derive recommendations for product design in a way that:

1. End-user musts and wants are identified and quantified for importance.
2. Company and other manufacturers product features are evaluated for current and future performance (competitive and performance analysis).
3. Product design features are correlated with marketing and business plans.
4. Appropriate technology or methodology is used to provide necessary product functions for accomplishing the user's task.
5. Appropriate manufacturing methods are utilized to build product.

These, in turn, are integrated in a matrix structure to assist one in concept selection so that:

1. Products satisfy user needs in performance, price, and delivery.
2. Products satisfy company needs and plans.

A customer purchases a product in order to accomplish some task with certain performance requirements. The product provides utility to the user; utility, in turn, is provided by product features and functions. The manufacturer, therefore, produces functions by designing the proper features and performance into a product that the customer is willing to purchase. Value is determined by how well the product allows the customer to perform the desired task for the price paid. In this respect, value is determined by the user, not the manufacturer. Selling price is directly related to user-perceived product utility and performance and is not necessarily related to manufacturing costs.[1] COPC is a method used to design a product with value added to both the user and the producer at a cost, quality, and price that is acceptable to both. The COPC process is especially useful for concepting consumer-type products, services, and systems.

THE PROCESS

The process about to be described is an integrated model that combines: (1) applied behavioral science and stake building philosophy to initialize the process, and (2) a matrix incorporating value measurement, function definition, competitive analysis, a function-technology morphology, and elements of QFD. The team deliverable resulting from using the COPC process is a series of recommendations (technology paths) representing different product concepts that have been screened by customer and manufacturing criteria. The process begins by the team addressing study purpose, decision maker, scope, study completion date, correct study team members, and assumptions. These issues are covered in Chapters 7 and 8 in more detail.

The team consists of members with diverse interdisciplinary backgrounds, such as marketing, sales, customer service, design engineering, manufacturing, and quality assurance. Acceptance and credibility of results can be affected by who is or is not on the team. Unfortunately, power and politics are real issues to contend with. It is recommended to use a core team of knowledgeable people throughout the project. Many problems can be encountered if the makeup of the group frequently changes. For example, considerable time can be lost just getting new members up to speed in both content and process. Also, ratings and evaluations will become much less consistent.

The COPC process is used to design one product. It is not used to develop a product strategy for a family of products. If the product serves many different customers, a separate COPC should be performed for each individual user. It is also based on the assumptions that: (1) the company or business unit strategy is clear, and (2) that the country, market, segment, buyer, and user have been clearly defined and documented by the business unit (a big assumption indeed!). If the above strategy and marketing documentation is not available or doesn't exist, the COPC team must develop the data themselves. If this is the case and the team has developed these data, they must be presented to upper management and the business unit for verification before progressing further in the process. Sufficient time must be allocated in the COPC process for verifying (or developing) country, market, segment, buyer, and user, because markets can be complex. Too many times answers to these items are taken for granted. This merely increases the chances of the COPC team designing the wrong product. Caution must be exercised on these topics because marketing and sales can be offended or embarrassed, especially if they really don't know the answers! A structured process like COPC surfaces many gaps in knowledge and information across many areas from distribution to manufacturing to marketing. This can, of course, be threatening to those who speak in mysteries with a high fog index and those whose career has been founded on deception and turf protection. Consequently, it is advantageous if the process facilitator is experienced in the behavioral science techniques discussed in Chapter 8, such as interpersonal skills, team building, and leading groups and workshops.

CONSTRUCTING THE COPC MATRIX

The matrix is composed of three parts: (1) product functions and components, (2) the marketing/customer evaluation, and (3) the manufacturing/technology evaluation. An example of a matrix using a videocassette recorder (VCR) as the product to be designed will be used (see Figure 4.1).

The following background information is offered to inform you more about the example product. Our product competes with all the myriad other VCRs on the market. Our model is used to record television signals as well as play back the programs. It is also widely used to play back other prerecorded videotapes, such as those rented or purchased at videotape rental stores. The following hypothetical marketing data are defined.

Country: U.S. only.
Market: Video.
Segment: Home entertainment.
User: Family.
Decision Makers: Parents.
Chief Buying Influence: Primary income producer.

CUSTOMER ORIENTED
PRODUCT CONCEPTING

PRODUCT: VCR
TASK: Play Video on TV
DELIVERABLE: Electronic (TV) Image
BASIC FUNCTION: Display Electronic Image

COUNTRY: U.S. Only
MARKET: Video
SEGMENT: Home Entertainment
USER: Family
DECISION MAKER: Primary Income Producer

MARKET/CUSTOMER												MANUFACTURING/DESIGN							
	CUSTOMER REQUIREMENTS		COMPETITIVE ANALYSIS						PLANNING			MANUFACTURING CRITERIA		TECHNOLOGIES (HOW TO)					
														Tech 1		Tech 2		Tech 3	
Operational Functions	Features	Importance	Current Us	OM1	OM2	OM3	Desired Us	Improve Ratio	Sales Point	Score	Percent Score	Mfg Criteria	Percent Weight	Mfg	Cust	Mfg	Cust	Mfg	Cust
Load Tape	Ease to Load	10	5	8	7	10	10	2.00	1.5	30.00	58.7	Cost	15	5	7	5	10	5	10
	Ease to Orient	10	9	9	9	8	9	1.00	1.2	12.00	23.5	Dev Time	10	3	9	5	9	5	10
	Instructions	8	7	8	8	8	8	1.14	1.0	9.12	17.8	Durability	15	5	5	5	6	5	9
										51.12	100.00	Time to Complete	5	1		5		4	
												Quality	20	5		4		4	
												Reliabilty	20	4		4		4	
												Maintainability	15	4		4		4	
													100	425	711	445	905	440	982
Transport Tape																			
Record Image																			
Playback Image																			
Rewind Tape																			
Remove Tape																			

Figure 4.1 Customer-oriented product concepting matrix (VCR example).

CUSTOMER-ORIENTED PRODUCT CONCEPTING PROCESS (COPC)

Part 1: Product Functions and Components

Step 1: Define Basic Function

Purpose: To establish the reason for existence of the product.

To begin building the matrix, the interdisciplinary team first defines the task that the user wants to accomplish by purchasing the product. To define the task or deliverable, the team members envision themselves as the user and ask why they are purchasing the product. For this example, the customer task is "play video tapes on the TV set." The team then defines the deliverable produced by the product, in our example, "an electronic (TV) image." Based on the user's task and product deliverable, define the overall basic function of the product. The basic function is the prime reason for existence of the product. It is what makes the product work or sell. It is what allows the customer to accomplish the task. As in traditional value engineering, the basic function is described in two words: one verb and one noun. Personal analogies are helpful in defining functions. That is, you actually pretend that you are the product itself and ask, I am a ______, what do I do? Why do I exist? The answer must be stated in the verb-noun format. In our example, the basic function of the VCR is "display (electronic) image." Some other examples of basic functions are: wire, transmit signal; screwdriver, transmit torque; wall panel, partition space.

Step 2: Define Operational Functions

Purpose: To establish the operational functions that must be performed to accomplish the basic function defined in Step 1.

The operational functions that must be performed to allow the customer to accomplish the task and obtain the deliverable are defined. The basic question to be answered is: What are all of the necessary operational functions that must be performed in order to accomplish the basic function (and allow the user to accomplish the task)? Operational functions are also defined in the verb-noun format. They are listed where practical in sequential order of operation, the way the product is produced and used today. This is equivalent to a linear time-sequenced function flow. This linear format is dissimilar to a traditional FAST diagram, where functions are ordered by cause and consequence and not by time sequence.

Operational functions are defined at the highest (macro) level of aggregation. Their number should be approximately 6 to 12. Personal analogies are again helpful in operational function definition. The questions asked are: I am a ______. My basic function is ______. How do I ______? In order to permit my customer to accomplish this task, I must *(function), (function), (function), . . . , (function).* Operational functions are listed as the product performs or operates today. This function flow may not be valid for future designs, which may use technologies different from today's. Worry about that later. This is why you are building the matrix. In our example, the operational functions are load tape, transport tape, record image, play back image, rewind tape, and remove tape. Note that all functions have been listed. We included the function "record image" to account for the customer task of recording something. This function would not be used by the customer if the VCR is used only in a playback mode. However, we include it because the customer will also be recording TV shows. The basic function, however, is still display (electronic) image. A secondary function happens to be "record image" and this is one way of acquiring an image to display. The other way to acquire an image is to rent or purchase a prerecorded videotape.

Listing operational functions works best with processes, machines, procedures, and services. For products like packaging, consumables, durable goods, and so on, it is easier to list major components first and then convert the components to functions at a later time. For example, a 35-mm film package might be listed as case, box, plastic can, can lid, case graphics, box graphics, can label, can label graphics, carton physical characteristics, or instructions. Underneath the component name would be listed its function. Sometimes two separate columns are used in the matrix, one for components, the other for functions. Listing the function is especially important for the creativity phase, where the team brainstorms to function and not components. Determining the basic and operational functions is not so simple a task for the team. Function definition is a new language to which participants must become accustomed. The function definition provides the structure for completing the matrix. It acclimates the team members and provides focus to concentrate on what the product must or must not do.

Part 2: Marketing/Customer Evaluation

Step 3: List Customer Features by Function

Purpose: To list for each function all of the features and wants that the customer may encounter through the use of the product.

Once all functions have been defined and sequenced, Part 2 of the COPC matrix is started by listing for each individual function all of the features or customer requirements pertaining to that function. In our example, some, but not all, of the features of the function "load tape" might be "easy to load," "easy to orient cassette," "clear definitions," and "able to load in low light." Direct quotes from customers are great.

Customer features can be obtained from marketing intelligence, research data, focus groups, surveys, interviews, and internal company sources. The ideal source would be to invite and pay customers to be ad hoc members of the team! Typical categories of features are ease or difficulty of use, convenience, dependability, availability, appearance, information/instructions, maintenance, and service. Tom Snodgrass of the University of Wisconsin, at Madison, and president of Value Standards, Inc., has developed four categories of functions that have been used with considerable success in his *customer-oriented product design* method.[2] They are: assure dependability, assure convenience, satisfy user, and attract user. These categories are also helpful as stimulators for thinking about features.

Some functions may be invisible to the user; consequently these invisible functions may not have customer features listed in their row. Examples of invisible functions for a VCR are open cassette, thread tape, read magnetics, acquire signal transmit signal, and so on. What many teams do in these cases is list all customer needs that can be impacted by these functions. In this case, technology selection will be based on most favorable impact on customer needs.

It is best to record the features in the customer's own words when possible. If customer quotes are used, it is also recommended that a glossary of terms be established that includes precise translations of the features into terms that are more meaningful to the team and technical community. For example, customers might say they "want a sharp image." This is important to know, but the technical community needs to know how sharp is sharp. The descriptor would be recorded both on the COPC matrix and in the glossary of terms along with a descriptive translation. The feature as translated in the glossary might read: The image on the TV screen should have clear definition around the edges. Perception of sharpness may be influenced by contrast." The glossary is especially useful if there is turnover in team members or if there are ad hoc members. It helps get members up to speed.

The question of what to do with things like government regulations/standards, international standards, and other regulatory requirements arises quite often. These standards are more than customer wants. They are givens and absolutes. They cannot be overlooked. To always keep them visible one can add them as a grouping in a separate row at the bottom of the list of customer functions and needs. The functional equivalent can be labeled as "satisfy regulations." There is no compromise with regulatory requirements. They are not brainstormed or tested, but they need to be satisfied. In addition to scoring technologies against customer needs and manufacturing criteria, they must be checked to see if they fulfill 100% of the regulatory requirements. This is equivalent to checking technologies for compliance with customer *musts.* Technologies are not rated against these standards. Instead, it is a dichotomous decision where compliance is labeled either "yes" or "no." Those technologies that do not satisfy regulations are either eliminated from consideration or modified until they do fit.

Step 4: Quantify Customer Features for Importance

Purpose: To determine how important each feature is to the customer.

After listing all of the features for all functions, the team must quantify them for perceived importance to the customer. One method that can be used is to establish a simple rating scale, say, 1 to 10. If a scale is used, descriptors should be written by the team for at least three or four anchor points in the scale (e.g., 1, 5, 10). The descriptors should, as best as possible, be stated in the words of the customer. Such a rating scale might be as follows:

Rating	Descriptor
10	I must have this feature. I expect it and would definitely switch brands and pay more to get it.
8	I would like to have it: all else being equal, I would probably change brands or pay more to get it.
5	It would be nice to have; it would make me happy, but I wouldn't go out of my way to get it. I might switch brands to get it but would not pay more to get it.
1	I am apathetic about this feature; it really doesn't influence my buying decision.

Descriptors written by the team members are used in order to promote consistency across raters. For example, without scale descriptors, what one person rates an 8, another person may rate a 10, and so on. Developing the descriptors encourages a dialogue among team members that results not only in more consistent ratings, but in a sharing of viewpoints that might not otherwise have surfaced. In turn, this dialogue leads to better understanding. Caution must be observed, because there is a tendency when establishing rating descriptors to state them more in marketing terms like "must have this feature in the product to have a superior competitive advantage." This is a producer-oriented phrase. The actual user of the product could probably care less about whether or not the company's product has a competitive advantage to the producer. The user is concerned about his or her own competitive advantage. Rating scale descriptors must be tailor-made by the team for each project/product. There is *no* universal rating scale!

The actual rating should be discussed verbally as a group. Numerical flash cards or an electronic voting system can be used to get discussion started. The full-scale range of rating numbers (1, 2, 3, 4, . . . , 10) is used for rating. It is advisable to have a team recorder/secretary record pertinent comments resulting from team discussions that lead to the final team rating. For example, it is not uncommon to have a bimodel or even a rectangular distribution of ratings. When this happens, the facilitator asks those who had high ratings to share their viewpoints leading to their ratings. Likewise, those who had low scores share their views. After the ratings discussion, the team is usually asked to vote again. The discussions and feedback generally lead to a convergence of votes in the second round. Considerable information is generated during these rating discussions that can be valuable later on. The notes taken document the rationale behind a rating. This is especially important months after the team sessions when memories start to fade.

What happens if the study team wants to list more than one customer for determining importance of customer needs? Will all customers have the same needs? Will all customers have their own measures of importance for each need? Our experience shows that the labels on the customer needs generally do not change across customers. What does change across customers is the importance of those needs. One way to capture all of this is to have an importance column for each customer. A separate importance rating would be given to each. In addition, weighting factors could be assigned to each customer according to that customer's importance to the company, volume of business, and so on. To do this, instead of using a rating scale 100 points would be allocated across the customers. The weighting factors would then be multiplied by the importance ratings. The products would be summed for each customer need row. The final sum of products would then represent an overall weighted importance across all customer columns. It is then possible to interpret the customer matrix to better understand the weighted importance scores. Caution is advised if you include a large number of customers, say more than 10. Including every possible customer can dilute the overall weighted importance score so that it can be almost meaningless.

It is obvious that marketing, customer representatives, or service and repair people should participate in rating the features for importance. The ideal method for determining customer importance would be to interview real customers or have them participate in the team exercise. This could prove difficult, however, raising issues of confidentiality and product disclosure.

Having the right people participate in this exercise enhances the credibility of the COPC project. Another advantage resulting from customer feature identification and quantification is that gaps or missing information readily surface. If answers are missing, question marks are inserted, and the topic is addressed later. The gaps in information can be very useful input for designing surveys or focus groups to obtain the needed information. Therefore, identified missing information can be just as useful and important as known information. Information gaps that surface highlight the fact that "we didn't know that we didn't know the information was missing."

Step 4a: Optional Rating Method. An optional and often-used method for rating importance is direct magnitude estimation.[3] It entails locating the most important feature in the feature group being considered and giving it the highest rating, say, a 10 (assuming a 1 to 10 scale is used). All other features are rated relative to this feature. For example, a rating of 3 would be about one-third the importance of a 10. It is possible and legitimate for other features to also have a rating of 10. Regardless of which rating method is used, it is important that the ratings reflect only one customer/user. One set of feature ratings can not be used for a family of products representing different users. If this is the case, separate importance ratings must be performed for each individual user. Very often the majority of the ratings are the same, but it is still important that every rating be considered. Scale descriptors should also be checked for appropriateness for rating each user or product combination. It may also be necessary to add, delete, or change feature descriptions.

Step 4b: Special Case. Experience has shown that, in some cases when the team is rating customer features for importance, they may discover that the features are not at the same level for analysis. There is a hierarchy of importance. This happens when functions or components differ in their level of indenture or when some features cluster as subsets under another feature. Consider an automobile, for example. A feature such as "withstand force" most likely would get a high rating, say, a 10, regardless of the function or component to which it pertains. However, a score of 10 for the feature pertaining to the bumper or seat-belt system is considerably different than a score of 10 for the same feature pertaining to the arm rest on the door. When this situation occurs, and it often does, the analytical hierarchy method,[4] also known as the *nested hierarchy method,* of scoring should be used to mathematically convert all feature importance ratings to the same equivalent level of indenture. This produces a treelike hierarchy whose number of levels of features may vary from two to four. Our VCR example has two levels, one level being functions, the other level being features. It may happen that the first level functions may vary in their relative importance. All of these functions are important, or the product would not work. However, some of the functions, relative to one another, are more important than others. When this happens, the features relating to the more important functions also take on a more important significance than those features belonging to the less important functions. So, even though some features may have the same second level rating, they may not in actuality be of equal final importance. The final importance rating is dependently factored by the relative importance of the function to which it pertains.

The following procedure (a two-level tree) is used to rate and stratify two levels of functions and features (see Table 4.1).

1. Rate the highest level functions (or components) for relative importance using a rating scale established by the team, as discussed earlier.
2. Normalize the function ratings so that the normalized ratings sum to 100. To do this, sum up the individual rating scores from Step 1 and divide this sum total into each individual function rating. The result is original rating scores converted to a percentage that sums to unity.
3. Repeat Step 1 by rating the features (second level) for each function.
4. Normalize the second level features within each function such that the normalized second level ratings sum to 100 for each function.
5. To obtain the final feature importance score, multiply the normalized feature scores (level 2) within that function by the corresponding normalized function score (level 1). The resulting product is the weighted feature score for each feature. By weighting the individual normalized

TABLE 4.1 Nested Hierarchy of Feature Importance Ratings (Two-Level Tree)

Function (Level 1)		Feature (Level 2)		Weighted Importance Score				
Score	%	Score	%	%L1		%L2		x 100
		10	50	0.32	x	0.50	=	16.0
8	32	5	25	0.32	x	0.25	=	8.0
		5	25	0.32	x	0.25	=	8.0
		Sum 20	Sum 100					
		8	80	0.40	x	0.80	=	32.0
10	40							
		2	20	0.40	x	0.20	=	8.0
		Sum 10	Sum 100					
2	8	Sum 8	Sum 100	0.08	x	1.00	=	8.0
		5	18	0.20	x	0.18	=	3.6
5	20	5	18	0.20	x	0.18	=	3.6
		10	36	0.20	x	0.36	=	7.2
		8	29	0.20	x	0.29	=	5.8
Sum 25	Sum 100	Sum 20	Sum 100					

feature score by the normalized function score, we have now brought the final feature scores to the same level of comparison. Table 4.1 illustrates the two-level nested hierarchy computation. The final scores are multiplied by 100 to make the final numbers more convenient. The final weighted feature importance scores derived in Step 4b are posted to the COPC matrix instead of the original nonfactored scores.

Sometimes situations require a three- or four-level tree. This usually happens when some features are subsets of other features, which, in turn, are nested under functions with different importance. A three-level tree involves the following steps (see Table 4.2):

1. Rate and normalize level 1 functions.
2. Rate and normalize level 2 features.
3. Rate and normalize level 3 features.
4. Compute the final feature importance score by multiplying the normalized function score (level 1) times the normalized feature score (level 2) times the normalized feature score (level 3). The final product of the three normalized scores is the final feature importance score calibrated to the equivalent level for analysis. Table 4.2 depicts a three-level nested hierarchy tree. Likewise, as with the two-level tree, the final factored feature importance scores are posted to the COPC matrix in place of the original nonfactored scores. In both cases, the hierarchy trees and calculations are retained for reference in case questions arise concerning their derivation.

TABLE 4.2 Nested Hierarchy of Feature Importance Ratings (Three-Level Tree)

Function (Level 1)		Feature (Level 2)		Feature (Level 3)		Weighted Importance Score
Score	%	Score	%	Score	%	%L1 x %L2 x %L3 = 100
		T		10	.40	0.32 x 0.50 x 0.40 = 6.4
		10	50	10	.40	0.32 x 0.50 x 0.40 = 6.4
				5	.20	0.32 x 0.50 x 0.20 = 3.2
				Sum 25	100	
				9	60	0.32 x 0.25 x 0.60 = 4.8
8	32	5	25	6	40	0.32 x 0.25 x 0.40 = 3.2
				Sum 15	100	
				10	59	0.32 x 0.25 x 0.59 = 4.7
		5	25	2	12	0.32 x 0.25 x 0.12 = 1.0
				5	29	0.32 x 0.25 x 0.29 = 2.3
		Sum 20	Sum 100	Sum 17	Sum 100	
				9	60	0.40 x 0.80 x 0.60 = 19.2
		8	80			
				6	40	0.40 x 0.80 x 0.40 = 12.8
10	40			Sum 15	100	
		2	20	10	100	0.40 x 0.20 x 1.00 = 8.0
		Sum 10	Sum 100	Sum 10	Sum 100	
2	8		(...)		(...)	(...)
5	20		(...)		(...)	(...)
Sum 25	Sum 100					

Step 5: Conduct Competitive Analysis

Purpose

1. To quantify how well current company product satisfies each feature (current state).
2. To quantify how well competitor's products satisfy each feature (current state).
3. To quantify how well company product should satisfy features in the future (future state).

After developing and quantifying all features and functions for customer importance, a competitive analysis is conducted regarding the performance of the company's current product, and other manufacturers, against each feature. The importance of the feature to the customer must not enter into this comparison. How important a feature is has nothing to do with how the products really perform.

To begin the analysis, a simple rating scale is used wherein the basic question being asked is: How well does this current product satisfy this feature? A rating scale is established to estimate how well products satisfy each feature. A typical scale is 1 to 10. Descriptors are written by the team for at least three numerical anchor points. A satisfaction rating might be as follows:

Rating	Descriptor
10	Fully meets customer need. In some cases may even exceed expectations. Gee whiz!
5	Satisfactory; not exceptional; not a problem or concern. Okay.
1	Unsatisfying; causes aggravation; customer may switch brands to avoid/eliminate this problem. Major problems.

As before, the scale and descriptors must be developed by the team so they have a scale that is meaningful and comfortable to them. The above scale is for example only.

All major competitors are listed. Sometimes there are too many competitors to list them all. In this case only the most important are selected. Using the full rating scale range (e.g., 1, 2, 3, 4, . . . , 10), the current company product and the competitors' products are scored for performance or fulfillment of each feature. If possible and where practical, real products should be available for comparison and evaluation. Scoring can be started with numerical flash cards or an electronic voting apparatus. It is important that team scores be discussed, especially if there is wide variation in scores.

Scoring products according to feature provides a competition benchmark for the current company product. This benchmark is next used to rescore where the company (team) desires to be, feature by feature, at some date in the future. The same rating scale is used. The future time horizon is established by the team and heavily influenced by marketing. This will usually be the product introduction date. Future-state rating should reflect the company's current standing versus competitors' products. In this case the perceived importance of the feature to the customer is figured into the ratings. The basic question being asked when rating the company's future product performance is: Given our current product performance in relation to that of our competitors, and based on the importance of this feature to the customer, where do we desire to be in the future with respect to this feature?

When doing the competitive analysis and rating the product's current and future performance, only one feature is considered at a time. That is, the company's current product is rated for performance on one particular feature. Then, immediately, the other competitors' products are rated for that same feature while thoughts and discussion are still fresh in team member's minds. After rating the company's and the competitors' current products for that particular feature, the team proceeds to rate company future desired performance.

Doing all of these ratings consecutively by row helps maintain the mind set and reference base for the feature. It is less confusing than rating one entire column (company) at a time across all features. When scoring where the company product is desired to be in the future, don't fall into the trap of thinking that there must be an improvement for every feature. The performance of many features will remain the same. Be realistic! There is a tendency to overachieve or overdesign. It is possible sometimes to have a performance rating greater than 10. This may set a new standard, or it may remain overkill. Here, as with feature importance ratings, recording of comments from team dialogue necessary to score the products is very valuable.

Rating desired company future performance is generally done by using the current company product as the reference. Sometimes this may not be so simple. For example, the current company product may be a manually operated product whereas all competitors' products are automatic. In this case, what is the reference base? What some teams have done in the situation is to do two comparisons to produce two desired future scores. The first comparison is with the company's current manual product. The second comparison is based on the average across all competitors' scores as the reference base. Both desired future scores can be useful. Reference with the current manually operated product gives an indication how much the company has to change based on current design and current manufacturing methods. The other score indicates how much the company has to change the product based on the average competitor's product. Both scores can be used to compute two improvement ratios, discussed next in Step 6.

Step 6: Calculate Improvement Ratio for Features

Purpose: To compute a ratio signifying how much change is desired for current company product for each feature.

In order to determine the amount of improvement needed for each feature, an improvement factor is computed. This can be done two ways. First, an improvement ratio can be calculated for each feature by dividing the desired future performance rating by the current performance rating. The resulting ratio is the amount of improvement needed for each feature. The ratio highlights those features needing attention and improvement in relation to customer need and competitive standing. There has been some controversy about computing a ratio. The argument has been, for example, that an improvement ratio of 9 divided by 3 is much more important than an improvement ratio of 3 divided by 1. The ratio does not really reflect where along the continuum of change and competitive position the ratio lies.

An alternative to the ratio is to use the arithmetic difference between desired future state and the current state. Using this method with the same above numerical example creates an improvement rating of $9 - 3 = 6$ and $3 - 1 = 2$, respectively. The feeling has been that the arithmetic difference maintains the focus of improvement and, in addition, places emphasis on the more important changes.

To avoid situations where the arithmetic difference might produce a zero, a 1 can be added to all resulting arithmetic differences. Thus, 1 + (future state − current state) = change factor.

Step 7: Quantify Market/Sales Point (Leverage)

Purpose: To quantify how much advantage marketing may have if changes are made for particular features.

Marketing and sales team members are asked for additional input called *market leverage* or *sales point.* Additional ad hoc marketing personnel may have to attend the team meetings to give additional input, or the marketing team member may canvass pertinent personnel outside of team meetings and bring the data to the team at a later date.

The basic question being asked of marketing for each feature is: Given the importance of this feature to the customer, and considering the amount of improvement (improvement ratio) needed, if we in fact make a change in this feature, can marketing take advantage of it? The questions are answered with numbers from a rating scale. The following sales point rating scale has been used:

Rating	Descriptor
1.5	Significant market leverage if change is made; most likely would mention it in advertising.
1.2	Some market leverage: technical representatives would have advantage on customer premises.
1.0	Status quo; no significant leverage; no negative impacts.

The ratings 1.5, 1.2, and 1.0 have been researched, studied, and used thoroughly over time. They seem to be empirical. These ratings have been carefully developed by the Japanese in their efforts to develop and apply the QFD process[5] discussed in Chapter 1. They provide sufficient information and, at the same time, lessen the impact of this rating on prior ratings of customer importance and improvement ratio.

Again, marketing must participate in this evaluation. If the team does this rating without marketing input, the COPC project will lose credibility. There have been cases where the sales point quantification step has not been used. Sometimes the customer importance rating, improvement ratio, or the competitive analysis is sufficient. To use or not to use sales point is a judgment call by the team on a per-case basis.

Step 8: Compute Feature Score

Purpose: To combine customer importance, improvement ratio, and market leverage into one figure to represent the merit of each feature for each function.

The feature score is a product of three numbers:

$$\text{(feature importance to customer)} \times \text{(improvement ratio)} \times \text{(sales point)} = \text{feature score}$$

After raw feature scores are computed for each feature, they are normalized to a percentage by dividing the total of all raw scores across all features for one function into each individual feature score. Note that normalizing is done separately for each individual function and is not based on the grand total of all scores across all functions.

The product of these three ratings, the feature score, draws attention to those customer features needing the most attention in designing a product based on customer needs. The purpose for normalizing is that these percentages will be used as a weighting factor for multiplying technology scores later in Part 3 of the matrix.

At this point in the matrix, we now have quantified customer features and functions. We have addressed the question: What are we trying to do for our customer? This is very useful information, but to stop here would be an incomplete analysis. What we have at this point is identify those features that will play an important role in designing the product from the customer's perspective. We now need input from the company and manufacturing.

Part 3: Manufacturing/Technology Evaluation

Step 9: Develop Technologies (Methodologies)

Purpose: To generate for each individual function a list of ways, methods, or technologies that might be used to accomplish or perform the functions.

It is now time to configure product concepts. To do this, we array against each operational function all of the known or possible ways of performing the function. This entails listing all of the technologies, methods, and procedures that might be used to construct a new product concept. One function is expanded at a time. Brainstorming or other creativity techniques are used to generate a list of all possible options that might be used. Each function will, in fact, require its own individual minicreativity session. The usual rules of creativity apply: go for quantity, no judging of ideas, hitchhiking is welcomed, and so on. The team is first encouraged to have a core dump of all of the obvious off-the-shelf ideas for achieving the functions. After this purge, the team then develops more blue sky ideas. Function-technology expansion is continued until the team feels they have saturated or exhausted the function-technology possibilities. The creative expansion process is done verbally as a group, so all team members benefit from interaction. It also encourages hitchhiking on other's ideas. Sketching ideas on chart pads also helps. Some teams have even added sketches on the COPC matrix.

It is during this technology expansion that the team finally gets into the nuts and bolts technology of building the product concept. It is now time for designers, manufacturing engineers, researchers, and so on to contribute their content knowledge to build a product concept. It has been suggested that the technology brainstorming be performed immediately after generating the operational functions in Step 2. The feeling is that doing the importance ratings, competitive analysis, improvement ratio, and sales point before the creativity session may inhibit the individual's creativity. The correct procedure is the one that the facilitator and team feel most comfortable with. Again, the COPC facilitator must have facilitating and interpersonal skills to handle these situations.

After this creativity step has been finished for all operational functions, the team will have created a function-technology morphological matrix.[6] This is a useful matrix. It can be considered a technology library cataloged by functions. It will provide the landmarks to chart a concept path to future design. The mechanisms for charting the path will be discussed next.

Step 10: Establish Manufacturing-Type Decision Criteria

Purpose: To document and quantify a set of manufacturing-type criteria to use for selecting technologies.

After expanding all the functions for technologies, it is now necessary to choose one technology for each function to create a technology path through the matrix. Each different combination of technologies creates a different path and thus creates a different concept.

How does one choose which technology to use for which function? It will eventually be done by scoring and weighting the technologies against two sets of criteria: one for the marketing/customer features, which we have already developed, quantified, and normalized, and one for manufacturing and design, which must now be developed.

A set of manufacturing/design decision criteria will be established for each function. Typical criteria are development cost, operating cost, installation cost, development time, time to completion, quality, reliability, maintainability, ease to automate, and quality-of-work life (QWL). After identifying and defining the criteria, they are weighted by the team for relative importance. This is done by allocating the team 100 points to distribute across the criteria such that the total sums to 100. This method is also referred to as sum-to-unity or constant sum method. The team does this by verbal interaction as opposed to individual distribution. The interaction necessary to reach group agreement is very information rich and worth the time. Sometimes the discussion can become very heated with strong arguments. This sum-to-unity method is more representative of the team's decision criteria than scoring each criterion individually on a scale (e.g., 1 to 100, 1 to 10) and then normalizing them to a percentage by dividing each individual criterion score by the sum of all scores. With the 100-point sum-to-unity method, the team is forced to deal with trade-offs across criteria; what they add to one criterion they must take away from another! These trade-offs give a more realistic picture of relative importance. All the criteria are important, but we want to know which are more important than others.

It is important that decision criteria be as mutually exclusive as possible in order to reduce confounding and interaction. For example, the criterion "waste" can be a subset of the criterion "operating cost," and using them both can give double accounting to operating cost.

These weighted criteria will be used as a set of weighting factors to multiply times technology scores to produce an overall weighted technology feature score that can be used for choosing appropriate technology. Next, how do we score the technologies?

Step 11: Score Technologies

Purpose: To quantify how well each technology satisfies: (1) the customer features, and (2) the manufacturing decision criteria.

Thus far in the process, two sets of criteria and their weighting factors have been established. They are customer features and manufacturing criteria. The next step is to rate each technology against both sets of criteria. A rating scale is established to allow team members to quantify how well the technology satisfies the manufacturing criteria and the customer features. The basic question asked is: How well does this technology satisfy each criterion? As seen before, a typical rating scale is 1 to 10. Descriptors are also written for several anchor points. A typical scale might be:

Rating	Descriptor
10	Satisfies feature/criterion in all respects. Ideal.
5	Satisfies feature or criterion. Okay.
1	Satisfies feature/criterion little or not at all.

The team decides which to rate first, customer features or manufacturing criteria. It is easier if the technologies are all rated against one set of criteria or features at a time. After scoring all technologies for customer features, the scoring process would then be repeated for the manufacturing criteria or vice versa.

This is less confusing than scoring the technologies for both customer features and manufacturing criteria at the same time. Also, when rating features and criteria, it is best to do so one row at a time. That is, rate all technologies for function 1 and feature 1, then rate them all for function 1 and feature 2, and so on. The same procedure would be used for rating against manufacturing criteria; function 1 and criterion 1, function 1 and criterion 2, and so on. Scoring one row at a time is much easier, because it provides a mental reference base that makes comparison easier. The same rating scale and descriptors are usually used to rate both sets of features and criteria, although separate scales could be established for each.

Step 12: Compute Technology Scores

Purpose: To calculate: (1) a customer feature score, and (2) a manufacturing criteria score for each technology to provide a basis for comparing the attributes and merits of technology options for each function.

At this point, each technology cell in the matrix has two sets of ratings, one for customer features and one for manufacturing criteria. These two sets of ratings will be multiplied by their respective weighting factors. Their products will be summed to produce two separate total weighted scores. That is, the customer feature score is obtained by summing the products of the individual normalized customer feature scores, developed in Step 8, and the technology feature score, developed in Step 11. Likewise, a manufacturing criteria score is calculated by summing the products of the individual manufacturing criteria weighting factors, developed in Step 10, and the technology manufacturing score, developed in Step 11. This process is repeated for each technology for each function row.

At this point there is a natural tendency to combine these two sums into one overall technology cell total score. The two summed scores for each technology should be kept separate. Combining them consolidates too much information. Important signals are easily missed. With one overall score, you are inclined to choose the technology with the highest combined score, which many times is not the best choice. Both scores are needed to search for tradeoffs and to balance the right combination of technologies for customer and company.

Step 13: Create Technology Paths

Purpose: To choose one technology in each function row to create a technology path representing the best combination of options for both customer and manufacturer.

The matrix has now been completed both qualitatively and quantitatively. The team is in a position to start choosing technologies to form a technology path across functions. The path is really a concept and not necessarily a detailed design. The path will be based on the best combination of function technology scores. Strong and weak points of each technology become obvious. The scores allow the team to discuss and choose the best combination of technologies that best satisfy both customer and manufacturer needs. Tradeoffs will have to be made. Each different combination of technologies represents a different concept. At first glance, you might expect that there are in infinite number of possible paths through the various combinations and permutations of technology cells. In reality, this is usually not the case, because there are combinations of technologies across function rows that are illogical or impossible for both customer or manufacturer criteria.

For example, sometimes it is electromechanically or physiochemically impossible to connect certain technology combinations. In any case, there are still many options available, and picking the best combination of technology scores will help surface options and identify the most likely candidates. To begin, several paths are created that have different objectives. For example, a least-cost path could be constructed by connecting the lowest-cost technology in each row regardless of whether they are logical or optimum combinations.

This path becomes the least costly concept that could ever be achieved and still meet customer and manufacture needs. It is highly likely that this path will be technologically impossible to achieve, but, if it were possible, it would be the lowest-cost alternative. The cost of this path then becomes the target cost. The same principle can be used to construct other paths for best quality, reliability, maintenance, QWL, and so on.

Selection of technologies may not be easy. Constructing the various paths begins to bound the various options. Some technologies will consistently surface for all paths. This does ease the selection process. Two graphic devices used to summarize the number of path options are the minimatrix and the strategy map.

Each column of a minimatrix represents a separate path (option). The actual names of the selected technologies are written in the minimatrix cells. The columns of the minimatrix show the boundary of path options. By scanning the function rows of the minimatrix one can easily locate technologies common across path options. Discovering such common technologies can prove very valuable. Figure 4.2 illustrates a minimatrix, a strategy matrix, of path options. Note for this example that for the function "provide resin" the technology "direct feed below hopper" is common to all options that have so far surfaced. Figure 4.3 shows yet another variation, a strategy map, for showing the interconnectedness of technologies and paths. Both devices show the same thing; some people may prefer matrix, others may prefer a map.

With some products, selection of one technology is confusing: What happens is that there is an interdependency among different function technologies. That is, whichever technology is selected for function x will constrain which technologies can be selected for functions y and z, and so on. Some teams select the technology for the most important function and let this selec-

	Technology Path Options			
Function	Least Cost	Best Overall Score	Best Precision & Reliability	Best Quality of Worklife
Prepare Dye	Today's Process	Pure Dye Solid	Pure Dye Solid	Pure Dye Solid
Provide Resin	Direct Feed Below Hopper	Direct Feed Below Hopper	Direct Feed Below Hopper	Direct Feed Below Hopper
Provide Dye	Manual Weigh	Manual Weigh	Manual Weigh	Valometric
Introduce & Mix Dye (Batch)	Today's Process	Mechanical Mix	Spray Coat	Batch Fluid

Figure 4.2 Strategy matrix (minimatrix paths).

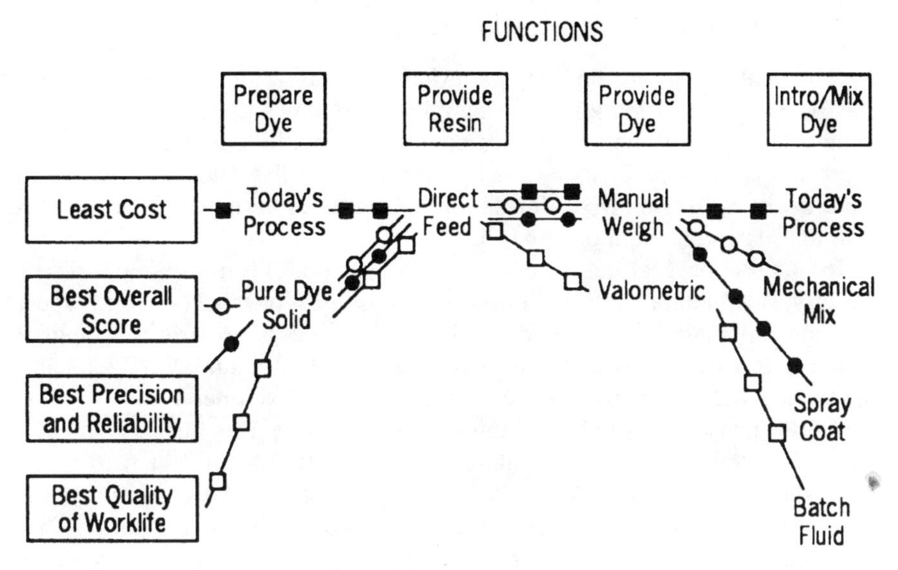

Figure 4.3 Strategy map.

tion determine the technology selections for the remaining functions. To do this, they assign importance weighting factors to all functions using the 100-point sum-to-unity rule. They then let the lighter weighted functions guide their selection of technologies.

Step 14: Select the Best Technology Path

Purpose: To select the path(s) that will become the final recommendation(s) for product design.

The team narrows down the options. After presentation and recommendations, the concepts are eventually turned over to design and manufacturing, who will start prototyping, testing, and designing. It is not uncommon that two paths are developed for final recommendation, one for "today" and one for "tomorrow." The today path is based on the best combination of technologies that fit into current time constraints to get the product out on schedule. The tomorrow path is clearly the best choice, but still requires some research or development that will not permit meeting current schedules. It is this path that sets future direction for marketing and manufacturing. It is also the concept that will bring the most added value to customer and company and that will allow best fit with the company strategic plan. All of the other least-cost and quality paths will still be retained for reference for presentation to management. Retaining these paths in an appendix will document and support current decisions. They can be used to show management that all possible options were considered in deriving the current recommended concepts.

Okay, You've Built the Matrix, Now What?

At this point, several technology paths have been developed. However, there may be gaps in information. For example, there may still be questions about some of the importance ratings or some parts of the competitive analysis. These kinds of questions frequently arise. So, it will be necessary to verify some of the COPC matrix information. Many times verification will require customer data collection of some sort.

The COPC process helps define areas of needed information and directs the surveys and data gathering into areas of greatest payback. In this respect, the data and signals developed in the COPC process provide the input to design better questions for focus groups and customer surveys. We recall a classic case where a focus group questionnaire was developed independent of the COPC project. At the last minute, a representative from business research decided to review the questionnaire with the COPC team. Much to everyone's shock and amazement, more than 50% of the focus group questions were inappropriate and way off target. Little would have been gained if the original set of questions had been used. After this embarrassing discovery, the market intelligence representative participated on the COPC team almost to project completion. The result was an almost completely revised set of focus group questions.

In some COPC applications, prototype products have been built and used at focus groups. Focus groups may involve videotaping customer reactions from behind a two-way mirror. By observing a focus session, a COPC team can obtain first-hand information on customer reactions to the prototypes and/or to a prepared set of questions. The data from the interviews, questionnaires, and focus groups are summarized. The summary information is mapped back into the original COPC matrix. Importance and/or competitive analysis ratings are revised and updated. New features or criteria are added as necessary. The technology scores are recalibrated to reflect the data updates. The original technology paths are checked for validity. New technology paths may also surface. The matrix is fine-tuned to the point where a decision can be made on the technology path(s) to recommend.

The matrix should be continually updated and considered a living document. If the matrix is properly updated, the COPC process is self-correcting. Feedback loops often require the team to go back and update prior data, which, in time, provide some missing links and course corrections.

CAN WE WIN?

The output from a COPC application is a product concept(s). This concept has been checked for its fit with customer and company needs. The concept represents the team's best estimate of what it will take to play the game. The following questions are appropriate to ask once a concept path has been established:

1. What will it take to play the game? Does this concept fit that profile?
2. Can we play? Do we want to play?
3. Can we win?
4. How do we compare to our competitors' designs? Can they win?
5. Why should a customer buy this product (concept) as currently perceived?
6. Mr. Manager, this is what it will take to play the game. Do you still want to play?

As I discuss further in Chapter 10, COPC is used at a tactical level (Figure 10.3) in the company and employs a macro level of detail. The above questions are also macro level and are used as a "quick hit" for middle/upper level managers to check their overall business decisions regarding the product at the very early stages of the product life cycle.

So often companies do not ask these question early on in the product life cycle. The reason is that they generally do not have a mechanism (like COPC) for doing so.

HOW DOES COPC RELATE TO THE BUSINESS CASE?

In Chapter 8, I discuss the need for a design team to have access to the company business plan when developing products. Because COPC is a design process whose output is a product concept, it should certainly connect with the business plan. The recommended product concept should be checked with the business plan to be sure both are congruent with each other. Such a comparison can work both ways: Does the concept fit the plan and does the plan fit the concept? Use of the COPC process can provoke the dialogue for challenging the focus of the business plan.

The technology path(s) information can be used as input to a decision and risk analysis (DRA) process for building a business case. Some of the technologies surfaced in the COPC process become candidates for research and development projects. These projects are generally approved based on their fit to a business case. DRA and portfolio analysis are quite often the processes used to derive the business case. The COPC process integrates very nicely with this type methodology. That is, the use of the COPC process results in a concept/design and the application of DRA can help one develop a business case based on the COPC design both favorably and unfavorably.

INTERFACES

A completed COPC is an extremely useful tool to show all of the interfaces in a system. The ordered set of functions with their corresponding technologies provide the basis for revealing these interactions. As such, a logical quality tool to use after the COPC matrix is completed is Failure Mode Effects Analysis (FMEA). Teams are split into work groups that have been clustered around particular functions or function-technology pairs. Team members are assigned according to their area of expertise. For example, software experts work with software related functions, electro-mechanical people are assigned to these areas, and so forth. Using this multiple work group parallel approach allowed the FMEA to be done in a short time. One or more consolidation meetings are necessary to coordinate any correlations or dependencies across functions. As a result the COPC team produced concepts that were quickly analyzed for failure possibilities. Solutions to the failures were surfaced very early on in the life cycle of the product.

Using these and other quality tools also provides the structure for Concurrent Engineering (CE) and co-location. In fact, it was during the FMEA activity that team members were all moved to the same section within the same building. Co-location made it easier for the team to do the FMEA and vice versa.

The manager of manufacturing for the project was also brought into the team at this time. This is CE in action! The function-technology matrix also facilitates the use of Design for Assembly (DFA) and Design for Manufacturability (DFM) processes. Again, this gets the manufacturing and assembly people involved earlier in the commercialization cycle.

People tend to underestimate the power and utility of developing and studying interfaces. The human factors people can also take advantage of a COPC matrix. The functions, the features, and the proposed technology are all on the same document, sharing the COPC matrix data with the ergonomics and industrial design people will also get them involved earlier in the commercialization process. One of the end results of the COPC, FMEA, DFA, DFM activity is reduced product cycle time.

GENERAL OBSERVATIONS OF THE PROCESS

The process of building the matrix is fairly straightforward. The behavioral and scoping activities to initiate the process are more complex. The perceived simplicity of the matrix can also be a detriment, because many teams believe they can run the process and build the matrix themselves. The entire process from launch to finish is best done with a neutral third-party facilitator to lead both team building and process. Otherwise, the team can easily get mired in turf protection, drown itself in details, get lost in the woods, and quickly rush to solve unclear problems and shortcut the process. The process facilitator provides the COPC model and keeps the team on track. The facilitator's function is to keep the team well focused and make decisions on how long to dwell on a particular problem. This person does not get involved in content discussion and ratings, but does ask devil's advocate questions while, at the same time, keeping the pace moving.

Meeting length and frequency are important. As with the other processes discussed in this book, the initial launch meeting is best scheduled for a full day. Regularly scheduled follow-up meetings should be four hours and held at least once per week (twice a week is better) until the COPC project is finished. Because of the complexity and detail in designing products, infrequent meetings less than four hours long do not work as well as longer duration meetings. Team members spend too much time getting back to speed to where they previously left off in the process.

Using the right number and blend of core team members is important for developing a well-rounded knowledge base and for minimizing the effects of bias and political influence. Keeping the same people on the core team throughout is very important. It takes time to develop team cohesiveness and a team benchmark/framework for making judgments. Excessive changes in core team membership fractures the team bond and the mental reference base used in decision making.

The credibility of the COPC process and output is highly dependent on the consistency of team decisions, ratings, and applications of the process. Group consensus is the preferred method of decision making. However, where reaching consensus may take considerable time, the process facilitator should guide the team to proceed with a majority decision and write a minority report. Good meeting notes should be compiled so that the reasons behind group decisions can be retraced if changes have to be made at a later date. Many facilitators maintain an ongoing list (referred to as a "parking lot" in Chapter 8) for recording ideas, questions, and action items that surface prematurely out of sequence in the COPC process. This is very useful; it helps bring focus on these issues later at a more appropriate time. The parking lot list of items is always kept visible at all team meetings, either on wall charts or typed notes.

The COPC matrix is to be considered a living document that is continually updated with both qualitative and quantitative information. Product variables, customer needs, environments, and technology change with time. The matrix should be updated accordingly. Building the matrix on an electronic spreadsheet greatly simplifies the updating process. We have even constructed matrices on-line during the team meetings using a laptop or tabletop PC. An electronic screen coupled to the PC to project the computer image on a room-size projection screen has also proven very useful.

With the matrix on a spreadsheet, the team now has the opportunity to more easily interact with other team members and decision makers in a simulation mode. Numbers can be changed and scores recomputed instantly in a "what if" capacity. This is very appealing, especially to decision makers, and such interaction serves as stake building to foster ownership in the process, which in turn increases chances for acceptance of recommendations.

An interesting variation in using the electronic spreadsheet is to use it with customers/users. The entire COPC matrix would be developed by the team and put on the spreadsheet. Customers would then be asked to interact with the matrix. They could verify the customer needs by entering their numbers for their importance of the needs. They could also add additional needs that have not been included in the team-generated matrix. After direct customer input, the program would be run to recalculate technology scores. The on-line customer interface is particularly useful for those kinds of products that must be tailor-made for specific customer needs. In this case, the technologies in the matrix may be various interchangeable modules that can be integrated to form various customized configurations. The customer's importance ratings would highlight those modules best suited for the customer application.

In this respect, the glossary of terms is very useful. It provides a consistent language in the midst of change and amendments. The scoring and rating processes require a good understanding of the item(s) under discussion. Experience shows that considerable time can be lost due to confusion when terms are not specifically defined. Good recorded definitions not only speed up the rating processes but promote better consistency as well. Communication is improved through a better shared understanding of information.

Believe it or not, information can sometimes be a barrier to initiating a COPC process. We refer to this as *information paralysis,* the condition that occurs when teams think they know nothing about customers and product performance parameters. They believe conducting customer surveys and collecting data have to be done before beginning a COPC project. The belief is usually unfounded, because it is highly unlikely that intelligent people involved in the sale of products know nothing about the use and performance of those products. The COPC process draws out the collective knowledge of the team and brings all team members to a higher level of knowledge.

FUNCTION CORRELATIONS

In Chapters 1 and 2, I described an intercorrelation roof on top of the house of quality (HOQ). This roof matrix allows the team to identify and quantify the impact, if any, that a change to one product/engineering characteristic may have on any others. So too, an intercorrelation matrix may be added to the left side of the COPC matrix. The intercorrelation can be used with the function rows (see Figure 4.4). Such a function roof can highlight any sensitive function intercorrelation. These intercorrelations may be helpful when considering various technology path options. They can help in many tradeoff decisions.

In addition to intercorrelations, functions may also be rated for importance by using any of the rating schemes discussed so far in this book. The two schemes can be used together to judge the importance or sensitivity of functions. If a function rates highest in importance and also affects many other functions, we now have another signal to watch for when choosing technology to deliver that function. Conversely, if a function rates low in importance but affects almost all other functions, we immediately have a signal that something does not seem quite right. The point is that function intercorrelations and the importance ratings put us in a position to ask some important questions that may never have been asked before.

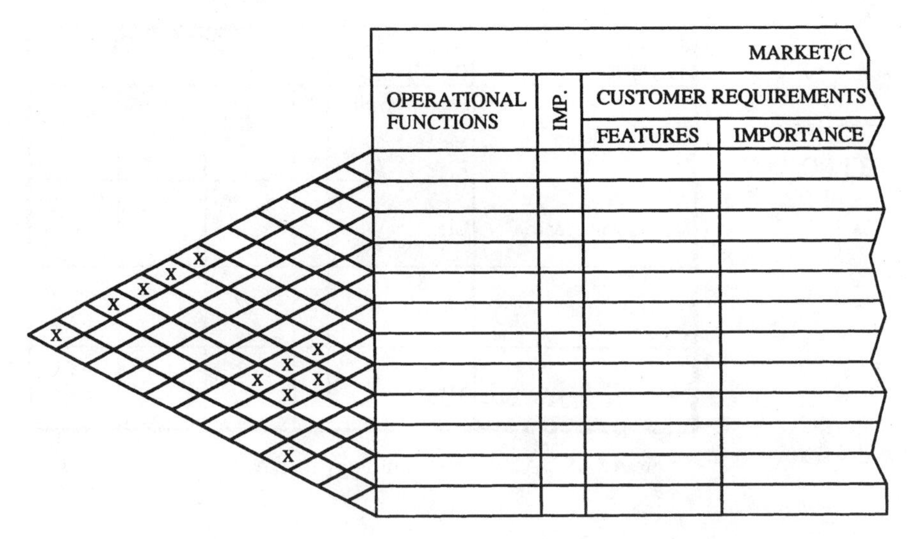

Figure 4.4 COPC matrix with function correlations.

There is considerable depth to a COPC matrix and all its various labels, tools, and ratings. COPC consists of tools within a tool. The depth of dialogue potential is directly proportional to the depth of the COPC matrix and its tools.

AN ALTERNATIVE METHOD FOR CONSTRUCTING A COPC MATRIX

Instead of building the COPC matrix as a single matrix, it can be split into two separate matrices, the technology matrix and the customer matrix. The technology matrix would contain the operational functions as explained and derived in Step 2, the manufacturing criteria (Step 10), and the brainstormed technology (Step 9). The criteria would be scored for importance (Step 10) and the technologies would be rated against the criteria (Step 11). Technology scores would be calculated (Step 12) and technology paths would be developed that best fit the manufacturing criteria (Step 13). You will notice that this matrix is identical to the right half of the regular COPC Matrix (Figure 4.1) as described earlier. Figure 4.5 illustrates the first matrix, the technology matrix.

In Matrix 2, the customer needs matrix, we introduce customer needs similar to the left side of the standard COPC Matrix (see Figure 4.6). However, this matrix will be constructed differently than previously described. The alternative method involves developing a modified Pugh matrix[7] wherein the complete technology paths from Matrix 1 will be rated against the customer needs. Selected paths from Matrix 1 are entered as column headings in the Pugh matrix. To estimate how a product (path) satisfies customer needs, it is better to use a holistic approach and consider the entire technology path. All operational functions do not perform in isolation in a product. We must consider the

COMPANY TECHNOLOGY FILTER

SUB-SYSTEM	COMPANY CRITERIA	I	TECHNOLOGY		
Manage Media	People Avail. Cost Risk				
Move Print Head	Etc.				
Control Print Head	Etc.				
Etc.	Etc.				

Figure 4.5 Alternate technology matrix.

interaction of all functions and technologies in concert. After all, this is how the customer will use them.

To construct the modified Pugh matrix we list the customer needs as one list rather than clustering them by operational functions. The team rates the customer needs for importance as in Step 4. The scores are then normalized so their sum equals 100%. This information is entered in the far left column. Next the reference or target product is listed in its own column. The next columns represent the selected paths from Matrix 1. For example, Path P1 could be the best quality/reliability path, P2 could be the least-cost path, P3 could be the best performance, and so on.

Two operations are now performed on the technology paths: (1) paths will be scored as to how well they satisfy the customer needs (similar to Step 11), and (2) paths will be rated as to how they compare to the chosen reference product (many times the competition).

First a satisfaction rating scale is derived for rating the paths for each customer need. Using this scale the paths are rated one row at a time. A path column score is derived by: (1) summing the raw column ratings, or (2) multiplying the normalized customer need importance times the corresponding path ratings and summing the products for a weighted column score. Which option to use is left to the reader. These totals will give the team an overall reference as to how well the paths fit or satisfy the customer needs. Individual cell scores will indicate any specific strengths or weaknesses of the paths.

The next operation is to compare the individual paths with the reference product. This is done by entering plus (+), minus (–), or zero ratings. If a path is better than the reference product for a specific customer need, a plus sign (+) is entered in the respective cell. If the path is worse than the reference, a (–) sign is entered and if it is equal to the reference a zero is entered.

Each path column is now summed for the total (+), (–), and (0) as well as the algebraic sum.

CUSTOMER
FILTER

CUSTOMER CRITERIA	IMP		TECHNOLOGY PATH					
			P 1		P 2		P 3	
Speed	15	REFERENCE	5	+	3	+	5	+
Quality	15		3	+	3	o	5	+
Reliability	20		3	o	4	o	4	o
Etc.	5		4	-	4	+	5	+
Etc.	5		2	o	3	-	4	+
Etc.	10		1	o	3	+	5	+
WGT. SCORE			220		235		285	
RAW SCORE			18		20		28	
Σ +				2		3		5
Σ -				1		1		-
Σ o				3		2		1
Sum				1		2		5

Figure 4.6 Modified Pugh (customer) matrix.

What the team now has is a reference document illustrating how well a particular path satisfies customer needs as well as how it compares to the reference, usually the competitor's, product.

Because the paths came from Matrix 1, we know the technology has already been studied in relation to the company needs. Matrix 2 incorporates the integrated technologies and allows comparison with customer needs and competitive position.

Earlier in Step 3, I discussed the issue of standards, regulations, and other dichotomous constraints. These regulations are grouped together and considered by themselves. With our current model of using the entire path we could construct a separate regulations matrix to work with regulations.

In a third matrix (Figure 4.7), all regulations would be listed in the far left column similar to the customer needs matrix above. Likewise, the paths would form the rest of the columns. No reference product would be listed. The paths would be checked to see if they comply (yes or no) with each individual regulation. Symbols, letters, or words can be used to indicate compliance or noncompliance.

STANDARDS, REGULATIONS	TECHNOLOGY PATH		
	P1	P2	P3
110/200 voltage	Y	Y	Y
Effluent	Y	N	Y
Ozone Emission	Y	Y	Y
Etc.	Y	N	Y
Etc.	Y	Y	Y
Etc.	Y	Y	Y
Etc.	Y	Y	Y
Σ Y	7	5	7
Σ N	0	2	0
FULL COMPLIANCE?	Y	N	Y

Figure 4.7 Regulations matrix.

In Chapters 6 and 9, I discuss the need for and some methods for looking into the future. One of those methods is a trend matrix (Figures 6.5 and 9.5). This simplistic device incorporates arrows to indicate the direction of a change. This method could also be used here to create a customer trend matrix (Figure 4.8). To do so the customer needs and their corresponding importance from Matrix 2 would form the left columns in Matrix 4. Two additional columns would be added representing two future time periods (in our

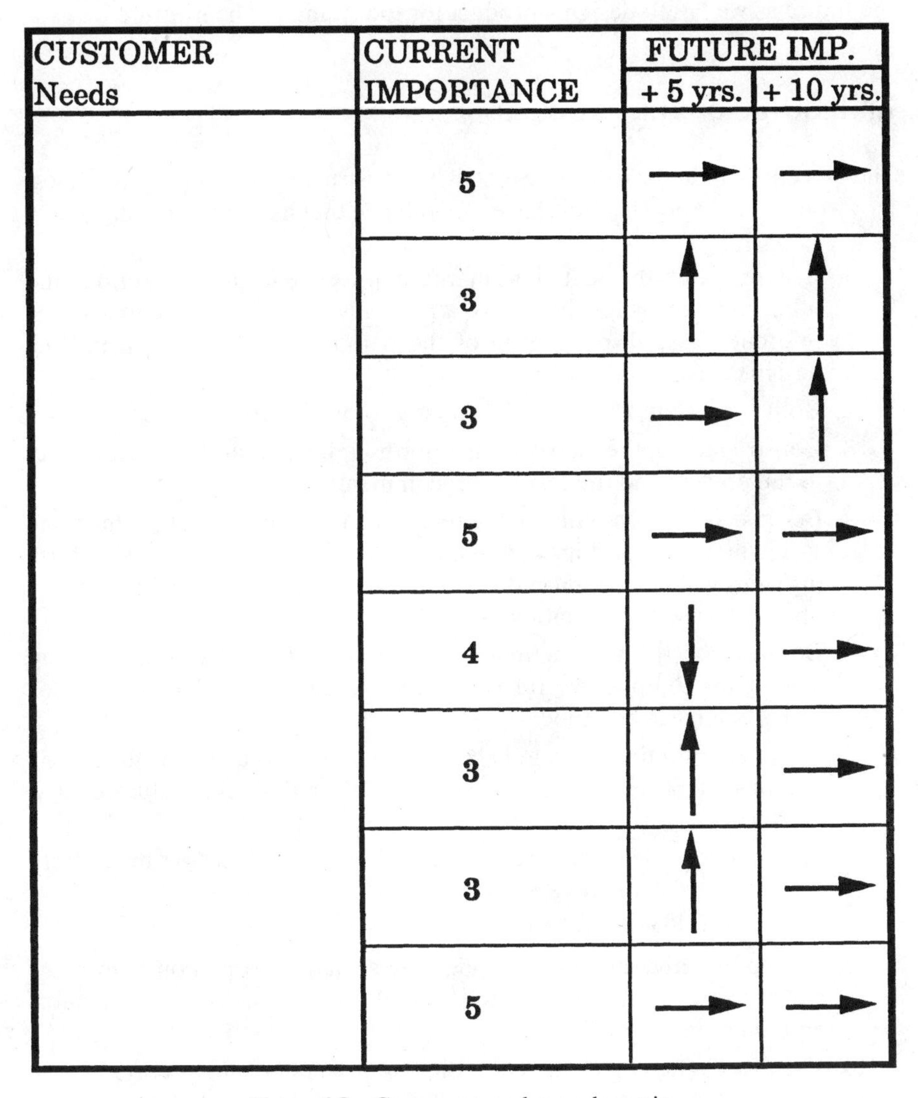

CUSTOMER Needs	CURRENT IMPORTANCE	FUTURE IMP.	
		+ 5 yrs.	+ 10 yrs.
	5	→	→
	3	↑	↑
	3	→	↑
	5	→	→
	4	↓	→
	3	↑	→
	3	↑	→
	5	→	→

Figure 4.8 Customer needs trend matrix.

example 5 years and 10 years) in the future. To indicate the trend in the importance of the customer needs the team uses arrows. An up arrow indicates the importance of the need will increase; a down arrow means the importance will decrease; a horizontal arrow means the importance will remain the same; and a zero means the need will disappear or no longer be important. If new perceived needs will come into being they should be added to the matrix. The time frame in this example, 5 years and 10 years, is completely arbitrary.

The four matrices may now be used together to set technology strategy for the future as well as to design a product for today and the near future.

ATTRIBUTES OF THE PROCESS

The process of initializing the matrix activity, constructing the matrix, and choosing a concept path provides a vehicle for dialogue that has many advantages.

1. It is probably the first time marketing, research, manufacturing, and other individuals involved have a mutually developed common-source reference base displaying all of the parameters necessary in making design decisions.
2. Both vertical and horizontal company communications are enhanced.
3. Construction of the matrix by an interdisciplinary team lends credibility to the process and the resulting recommendations.
4. Because it is a common-source reference base, the matrix permits one to see interrelationships and highlights information gaps. This, in turn, provides valuable input for designing surveys and focus groups to obtain the right information.
5. Because constructing the matrix is a structured process, it becomes an assault on complexity and forces participants to deal with complex issues one piece at a time.
6. When kept up-to-date, and because it is based on functions, the matrix will provide a technology reference base for designing future similar products.
7. The matrix provides the basis for designing a product that meets both customer and company needs.
8. The process highlights features important to the customer.
9. The competitive analysis provides the structure to pinpoint strengths and weaknesses for both company and competitors' product features and functions.
10. The process highlights areas needing improvement and attention.
11. Because it is function based, the process of expanding the technologies allows for greater use of creativity.

12. The two sets of criteria and the technology ratings allow you to choose the best combination of technologies to satisfy both customer and company.
13. Quantifying the matrix also promotes communication because numbers are easily debated and they highlight areas of disagreement.

COMPARISON OF COPC AND QFD

In case there is any confusion about what the difference is between COPC and QFD and how they relate to each other, I have assembled a comparison chart (Figure 4.9). Also in Figure 4.10, I have prepared a schematic COPC matrix that contains, for each section, the questions we are trying to answer and the respective output.

CONCLUSIONS

The COPC process has proven to be successful in a wide variety of applications. The reason for many of the successes is that they were also designed and planned politically and had top-down management support and funding. The process as designed and initially presented to management and sponsors had the following attributes that contributed to successful application:

1. There was a perceived advantage to using the process.
2. It was compatible with the existing company/department structure and operating philosophy.
3. The process is based on the existing language and vernacular of the organizations. There were no new threatening terms or fads.
4. The process does not have a high price in terms of anxiety, emotion, and comfort level. It appeared "safe."
5. There is little to lose if the process is terminated before closure.

Finally, the COPC process and QFD are both vehicles for dialogue. To be able to communicate, people and teams need a structure and a language. The QFD processes, like COPC, provide the structure through the matrices and the format/procedure used to construct them. A language is provided by the numbers that are derived and put into the matrices by the team members. Team members can debate different choices for numerical inputs. Considerable dialogue is required to merely derive the rating scales used to select numerical inputs to the matrix. The combination of the COPC matrix and the numerical language of value measurement provide the basis for enhanced communications. In this respect COPC/QFD will not make decisions but rather will structure input for users to make a better, more informed, decision.

COPC	QFD
1. COPC is perceived to be faster than QFD.	1. QFD is perceived to take a long time.
2. COPC produces a product concept.	2. QFD does not produce a design but rather a list of technical requirements.
3. COPC has a clear tangible end point (a concept).	3. QFD is complex and may have a confusing end point.
4. COPC is a design process.	4. QFD is a problem solving process (e.g., rusty car door), or sometimes a redesign process.
5. COPC focuses on what the product must do.	5. QFD focuses on technical parameters that are not necessarily focused back to a whole product.
6. With COPC, you know when you are finished.	6. QFD may not have a clear end point.
7. With COPC, you define the market, customer at the start of the process.	7. Market and customer are too often assumed and/or not clearly focused.
8. COPC works with the whole product versus pieces only.	8. QFD works with pieces (car door) versus the whole product.
9. COPC is an up-front macro level quick hit process.	9. QFD is a micro level drawn-out process.
10. COPC title is more understandable than QFD. Customer-oriented product concepting is explicit.	10. What is "quality function deployment?" This may take some explanation.
11. COPC does not take the product to the shop floor.	11. QFD will take you to the shop floor *if* all four matrices are used.
12. COPC results in a product concept that is used as input to start the QFD process.	12. QFD takes this concept in whole or in pieces to take it to the shop floor.
13. COPC is used in conjunction with company mission and business plan and uses VOC to help guide to a focused end point. It is outside-in holistically focused.	13. QFD starts in the middle of a problem and uses VOC for direction to a less focused end point. It is inside-out focused.

Figure 4.9 Comparison of COPC and QFD.

REFERENCES

1. Cook, T. F., "Determine Value Mismatch by Measuring User/Customer Attitudes," *Proceedings, Society of American Value Engineers* **21,** 145–156, 1986.
2. Snodgrass, T. J., and Kasi, M., *Function Analysis: The Stepping Stones to Good Value,* University of Wisconsin, Madison, 1986.
3. Meyer, D. M., "Direct Magnitude Estimation, A Method for Quantifying the Value Index," *Proceedings, Society of American Value Engineers* **6,** 293–298, May 1971.

1. What product? Evolutionary or Revolutionary? 2. What task is the customer trying to accomplish? 3. What is the basic function?	1. For whom are we designing product?

	FUNCTIONS	CUSTOMER	COMPANY
QUESTIONS	What are the FUNCTIONS necessary to provide the deliverable? What FUNCTIONS are needed to accomplish the BASIC function?	1. What does the customer want? 2. How important are they? 3. How does our current product satisfy customer needs? 4. How does our competitions product satisfy customer needs? 5. Where should we be in the future with product satisfaction? 6. Can marketing get leverage from the specified improvements?	1. What are all the possible technologies we night use to accomplish the function? 2. How well does the technology satisfy a) customer needs? b) company needs? 3. What are the company decision criteria used for choosing among technology? How important are they? 4. What are the most likely technology path options? What does that path look like? Will it win? Can it compete? Can we play the game?
OUTPUT	1. Function map/flow	1. Customer needs hierarchy 2. Competitive analysis 3. Target areas for improvement	1. Product concepts 2. Integrated technology path options 3. Weighted decision criteria 4. What it takes to play the game 5. Sourcing strategy

Figure 4.10 COPC question/output template.

4. Saaty, T. L., *Decision Making for Leaders,* RWS Publications, Pittsburgh, Pa., 1988.
5. King, R., *Better Designs in Half the Time, Implementing QFD Quality Function Deployment in America,* GOAL/QPC, Methuen, Mass., 1987.
6. Shillito, M. L., "Function Morphology," *Proceedings, Society of American Value Engineers* **20,** 119–125, 1985.
7. Pugh, S., *Total Design,* Addison-Wesley, Reading, Mass., 1990.

BIBLIOGRAPHY

1. Shillito, M. L., and DeMarle, D. J., *Value: Its Measurement, Design, and Management,* John Wiley & Sons, New York, 1992.
2. Shurig, R., "Morphology: A Tool for Exploring New Technology," *Long Range Planning,* **17** (3), 129–140, 1985.

CHAPTER 5

LINKING QFD TO PLANNING

BACKGROUND—CONNECTIONS AT THE TOP, PLANNING QFD (PQFD)

QFD is a planning and design process. As such it must be connected to the company mission and business plan. However, quite often, QFD is not checked against company plans or, too often, it is checked with the plans too late, after millions of dollars have been expended on a "nonproduct." There are huge amounts of prevented expenditures to be realized if only the QFD team could construct several pre-QFD or pre-HOQ matrices to bridge the gap between business plan and product design. Figure 5.1 illustrates a possible planning matrix chain from company mission to the HOQ or the COPC matrix. This model incorporates three planning matrices (P1, P2, and P3) to bridge the gap. I have called this process PQFD for planning QFD. This model is *not* meant to replace the corporate planning function! Based on QFD logic, the model takes current planning information and integrates it into a QFD/COPC format to make the needed connections to the QFD/COPC project.

The following PQFD matrices are discussed using a hypothetical example.

MATRIX P1—MISSION MATRIX

Matrix P1 arrays the elements of the company vision/mission with the main elements of the business plan (see Figure 5.2). The purpose of this matrix is to check if there are any mismatches between business plan and the company mission. The business plan elements are business unit needs at a macro level or the main elements in the plan. The business plan may also be known as a marketing plan or participation strategy. Each company has its own vernacular. The vision/mission should consist of only a few items, say, six or less. The dimensions of Matrix P1 should only be about 6 × 8. A relationship matrix is created using a ±1, ±3, ±9 rating scale. Notice that this scale goes in both a positive and negative direction. A possible impact scale

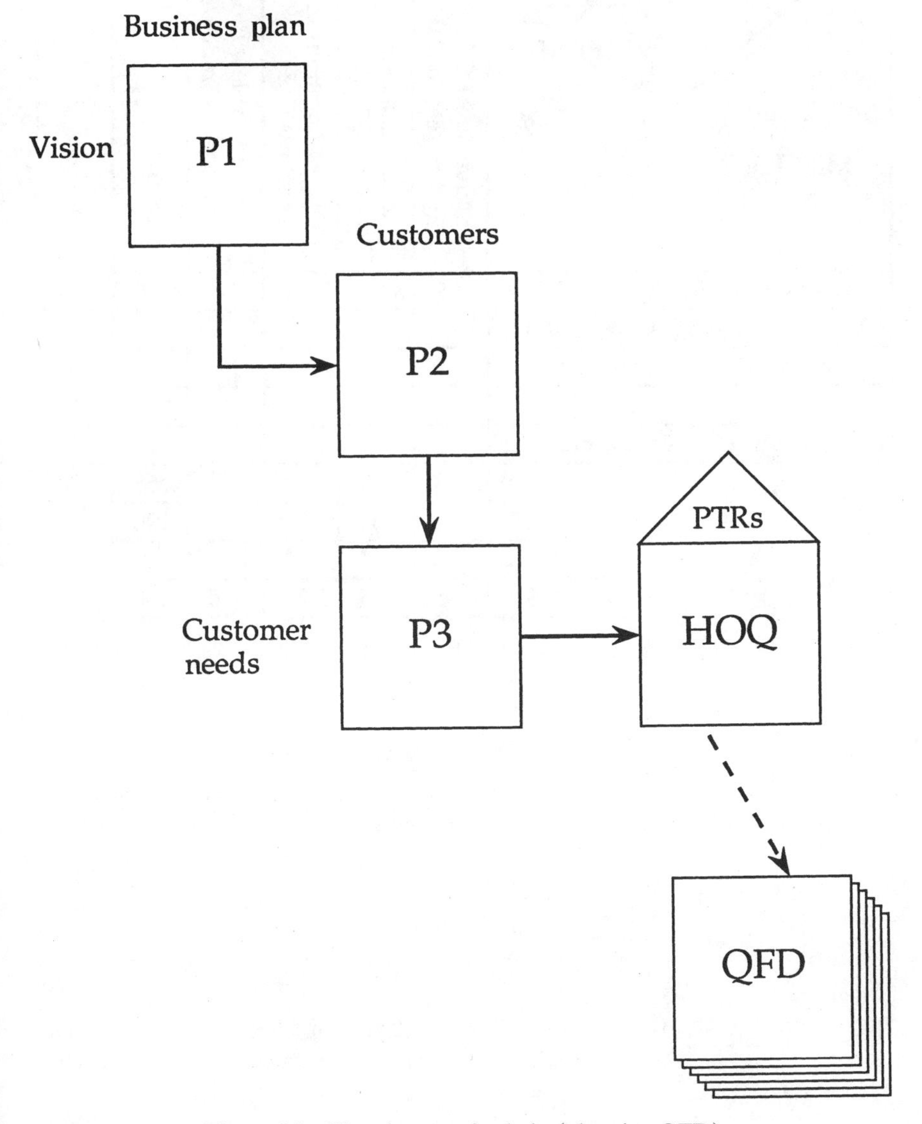

Figure 5.1 Planning matrix chain (planning QFD).

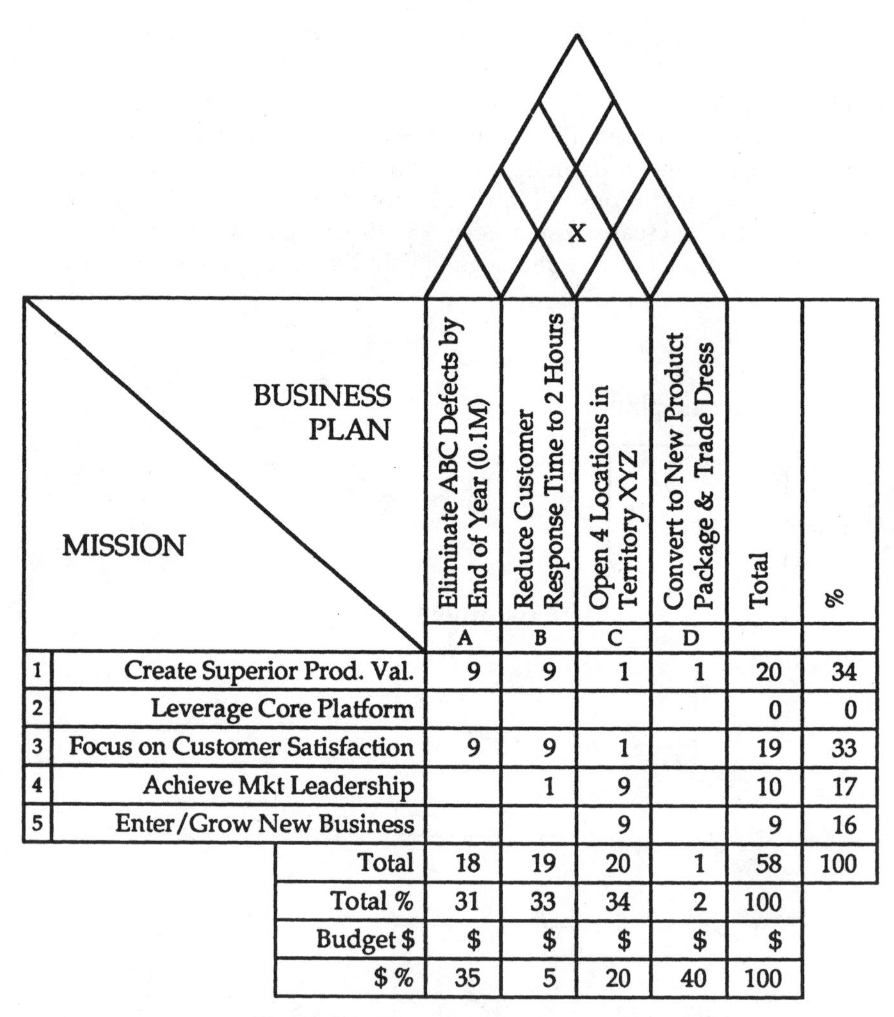

	BUSINESS PLAN / MISSION	Eliminate ABC Defects by End of Year (0.1M)	Reduce Customer Response Time to 2 Hours	Open 4 Locations in Territory XYZ	Convert to New Product Package & Trade Dress	Total	%
		A	B	C	D		
1	Create Superior Prod. Val.	9	9	1	1	20	34
2	Leverage Core Platform					0	0
3	Focus on Customer Satisfaction	9	9	1		19	33
4	Achieve Mkt Leadership		1	9		10	17
5	Enter/Grow New Business			9		9	16
	Total	18	19	20	1	58	100
	Total %	31	33	34	2	100	
	Budget $	$	$	$	$	$	
	$ %	35	5	20	40	100	

Figure 5.2 Matrix P1, mission matrix.

Mode	Scale	Descriptor
	+9	Business plan element will make it easier to accomplish the mission element; absolutely necessary.
Supporting	+3	
	+1	
Nonsupporting	0	Status quo; neutral; no influence.
	–1	
Defeating	–3	
	–9	Business plan elements will make it harder to accomplish the mission element; counterproductive; nonvalue-adding.

Figure 5.3 Impact scale, Matrix P1.

with descriptors is illustrated in Figure 5.3. The matrix relationships are derived by answering the following question: Does this current business plan element (column) make it easier or more difficult to accomplish the mission element (row)? What we are trying to determine is, if our company fully achieved the goals for the business plan, how would the mission statements be impacted? After quantifying the matrix, the column ratings are algebraically summed and normalized to a percentage if need be. Notice that it is possible that an algebraic sum may be negative. When normalizing the column totals, count any negative totals as zero. This will of course create a zero percent, which is sufficient for further study and analysis because the original impact figures are preserved in the impact matrix. These percentages now show a hierarchy across business plan elements in proportion to their impact on company mission.

Actual budget dollars can be entered at this point for each business plan element. These, too, are normalized to a percentage. This provides yet another perspective on resource allocation. For example, in Figure 5.2, we see that we are allocating 40% of our total business plan dollars for "new product package and trade dress." This business plan item has very little to do with accomplishing the corporate mission elements. How could this be? Can we justify it? Is there a logical explanation? Should we change it? These are questions to prompt a dialogue that needs to take place.

This same issue is nicely illustrated in a value graph (Figure 5.4) of the percent importance of the business plan element versus the corresponding budget dollars (percent dollars). Percent importance here refers to the normalized column totals. The position of item "D" on the plot graphically draws attention to some needed dialogue regarding the relationship of importance to cost.

Let us now use our data to compare the rows (mission elements) of Matrix P1. First, we sum the mission element relationship scores across the business plan element columns. This total is then normalized. These totals and/or percentages give us some indication of how much support the mission items are receiving from the business plan. If, in our example (Figure 5.2), we observe row number two ("Leverage the Core Platform"), we find that none of the business plan elements support this mission item. Here, again, it's time for questions and clarification. How can this be? Did we forget something? We also find that mission element number five is supported by only one business plan item. How does management feel about this? Based on the other business plan elements, every thing seems to be in order. But, at least we have an audit trail for a sanity check.

In traditional QFD fashion we can build a correlation roof atop Matrix P1 (Figure 5.2). Completing this matrix will allow us to check for any intersections, good or bad, among the business plan elements. In addition to having some impact on the corporate mission, they may have some impact on each other. If a business plan element does not support the mission, or even has a negative impact on mission, and in addition has a negative impact on one or

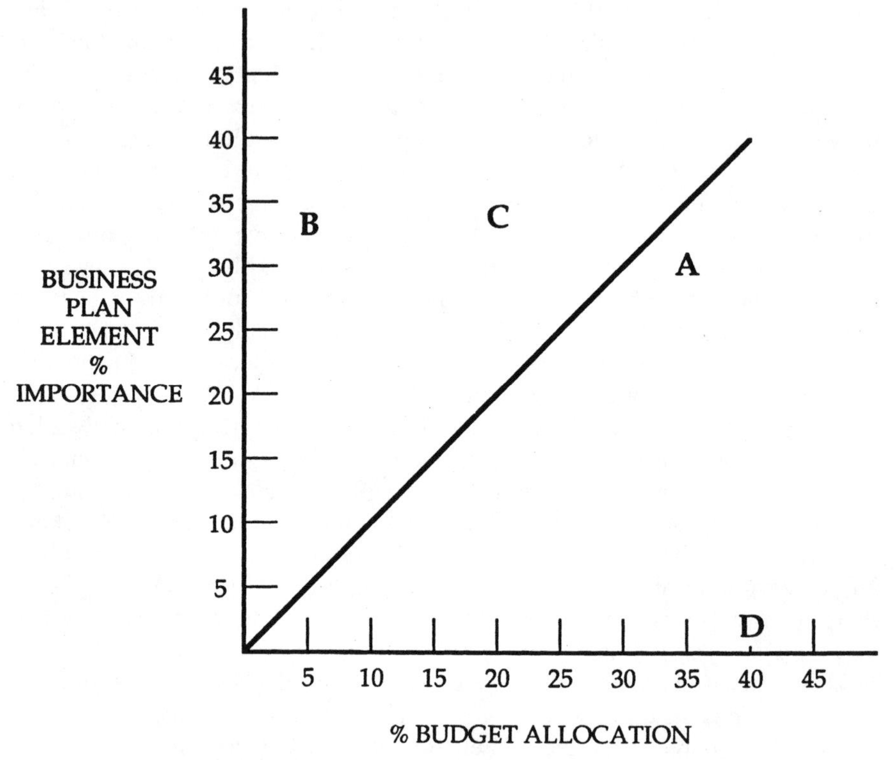

Figure 5.4 Matrix P1 plot, percent importance of business plan element versus percent budget allocation.

several other business plan elements, then someone should take a real hard look at this business plan element. How can it happen? What should we do? How did this escape our attention?

In our example, "Open 4 Locations in Territory XYZ" could have a positive effect and make it easier to accomplish the item "Reduce Customer Response Time to 2 Hours." This is in addition to all of its positive support to four out of five mission elements.

Who quantifies Matrix P1? The obvious choice is upper or middle management. A QFD facilitator can play a helpful role here. If the information about the mission and business plan can not be obtained, the QFD team could assemble and quantify Matrix P1 from their point of view. Their matrix would then be taken to management for validation or course correction. Another option is to have a QFD facilitator work with upper management to help them build and quantify the matrix. The QFD facilitator would then bring the completed Matrix P1 to the QFD project team to use as input to initiate the QFD study.

MATRIX P2—BUSINESS MATRIX

Depending on the size of Matrix P1, the business elements, or only the most important business elements, from that matrix become the rows of Matrix P2 where they are arrayed against the various customers within the market (see Figure 5.5). The purpose of this matrix is to check that we are addressing the right customer for our business plan or whether the business plan as written matches our current customer base.

A relationship matrix using the standard symbols for scores 9, 3, and 1 is quantified. The relationship numbers in this case represent how much the particular business plan element addresses, pertains to, or supports the specific customer. Is it possible that some of the relationship numbers could be negative? That is, are there some elements of the business plan that could cause aggravation to some customers? A positive and negative rating scale could be used here as well. A possible scale might look like that in Figure 5.6. The scaling is suggestive and is meant to spark your imagination to customize a scale that will fit your needs. Column totals are derived and converted to percentages. These figures give an indication of a customer support hierarchy. Both

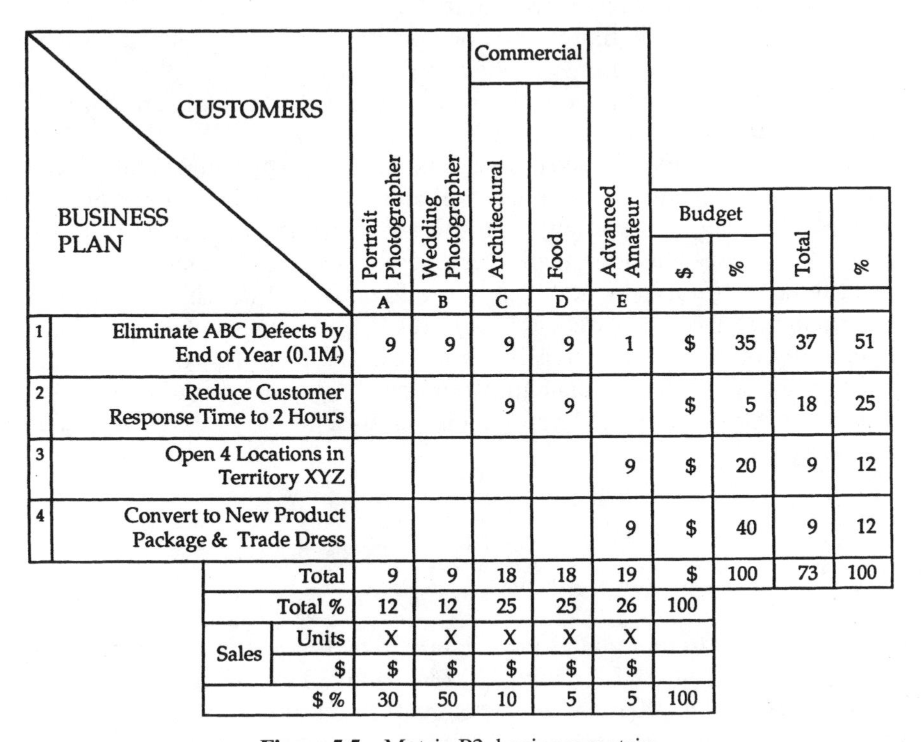

	CUSTOMERS / BUSINESS PLAN	Portrait Photographer	Wedding Photographer	Commercial: Architectural	Commercial: Food	Advanced Amateur	Budget $	Budget %	Total	%
		A	B	C	D	E				
1	Eliminate ABC Defects by End of Year (0.1M)	9	9	9	9	1	$	35	37	51
2	Reduce Customer Response Time to 2 Hours			9	9		$	5	18	25
3	Open 4 Locations in Territory XYZ					9	$	20	9	12
4	Convert to New Product Package & Trade Dress					9	$	40	9	12
	Total	9	9	18	18	19	$	100	73	100
	Total %	12	12	25	25	26	100			
Sales	Units	X	X	X	X	X				
Sales	$	$	$	$	$	$				
	$ %	30	50	10	5	5	100			

Figure 5.5 Matrix P2, business matrix.

Will the business plan element make it easier or more difficult to gain more of this customer's business?

+9	Business plan element will significantly increase our business with the customer.
+3	BP element will help grow business.
+1	BP element will attract customer's attention; may or may not get more business.
0	BP element will have no effect; status quo, business as usual.
−1	BP element will disappoint customer.
−3	BP may cause inconvenience.
−9	BP elements will cause customer aggravation; will open door for competitors to win over this customer.

Figure 5.6 Matrix P2, rating scale.

sales units and sales dollars are entered for each customer segment. Dollars are also normalized to a percent. We can now observe the relationship of sales dollars to customer support.

In our example 80% of sales came from customer areas that were addressed very little by the business plan. How is this interpreted? Does this indicate that our business plan should be reevaluated? Can we rationalize this discrepancy? This can also be vividly represented on a value graph as in Figure 5.7 where points "A" and "B" represent the above prose description.

Notice also that the business plan budget was carried forward from Matrix P1. The budget dollars were normalized.

Next the individual row scores are summed across customers. These totals are also normalized to a percentage. We can now make a comparison of business plan overall contribution to customers versus the budget dollars allocated for those business plan elements. Does the budget match the contribution? If not, why not? Can we rationalize and explain any possible discrepancies? Do a sanity check. These row figures can also be represented graphically on a value graph as in Figure 5.8. The point is, we are now in a position for some serious dialogue about customers and business plan. We took some existing data and put them into a structure to discover and study relationships.

MATRIX P3—CUSTOMER MATRIX

Matrix P3 is a customer matrix (see Figure 5.9). This type of matrix has been used for several years by many different companies. A customer matrix is always used when there are numerous customers and when you are trying to establish the importance of the many elements of the VOC. A separate

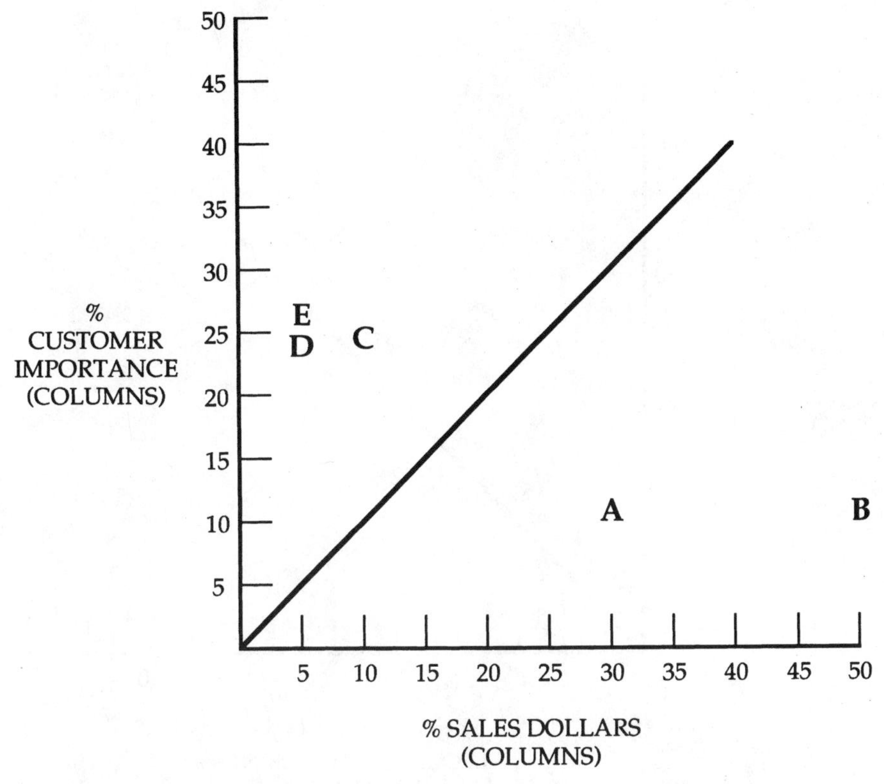

Figure 5.7 Matrix P2 plot, percent importance of customer versus percent sales dollars generated by customer.

importance rating for each VOC item is derived for each customer. The importance matrix allows us to scan importance of VOC items across customers. One can look for trends, similarities, patterns, clusters, and redundancies. Some teams have even averaged importance across all customers to derive one overall importance figure. Caution should be exercised in doing this, as it can give misleading results. Averages may also be normalized to a percent. A possible P3 scale might look like the following:

P3 RATING SCALE

Rating	Descriptor
9	Must have it
3	Nice to have
1	Ho-hum

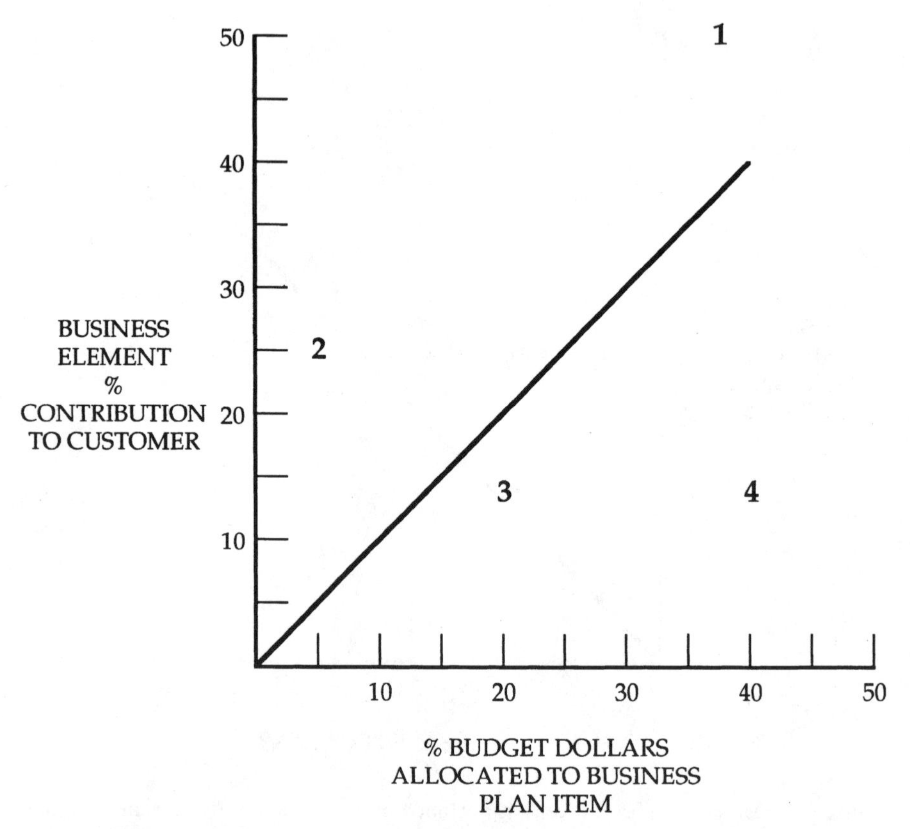

Figure 5.8 Matrix P2 plot, percent business element contribution to customers versus percent dollars allocated to business plan items.

Our Matrix P3 is really no different from other customer matrices. The only difference is how we arrived at which customers to include in the matrix. We actually derived our customers through the structure of Matrix P2.

We have to be careful in the use of Matrix P3. Remember, a customer matrix is used when there are many customers for whom we are trying to design a product. The problem arises when the team, marketing, or the company cannot decide for whom they are designing the product. There is a tendency to try to be all things to all people. Not only is this impractical but it is also poor business judgement. It leads to overdesign, unnecessary expense, and an unfocused product design aimed at too many users. The result: an overpriced short-lived product. Listing the customers from Matrix P2, or even listing all customers on Matrix P3, allows one to see commonalities, which in turn allows the team to develop clusters of customers. Such segmentation can allow the team to direct its efforts to the segment or cluster with the highest potential return.

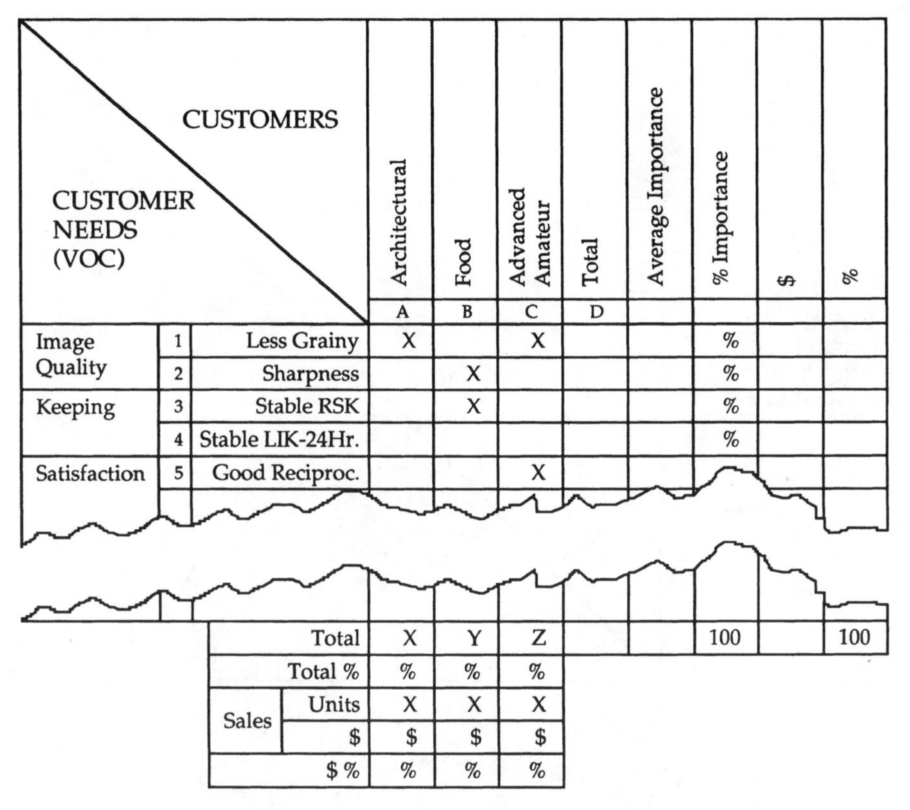

CUSTOMER NEEDS (VOC) / CUSTOMERS			Architectural	Food	Advanced Amateur	Total	Average Importance	% Importance	$	%
			A	B	C	D				
Image Quality	1	Less Grainy	X		X			%		
	2	Sharpness		X				%		
Keeping	3	Stable RSK		X				%		
	4	Stable LIK-24Hr.						%		
Satisfaction	5	Good Reciproc.			X					
		Total	X	Y	Z			100		100
		Total %	%	%	%					
Sales		Units	X	X	X					
		$	$	$	$					
		$ %	%	%	%					

Figure 5.9 Matrix P3, customer matrix.

We will add some other measures to the basement of Matrix P3. For each customer, enter the total yearly sales units and sales dollars. Normalize sales dollars to a percentage. When trying to determine what is important to whom, the relationship of sales volume to customer importance can encourage many questions. This relationship can also be plotted as in Figure 5.10. If many VOC elements are highly important to a particular customer from whom we get 40% of our sales, we just may pay attention to this fact. Conversely, if a customer has many important VOC items and has low sales, this can be a signal for market/customer development. The purpose of the matrix is to determine what is important to whom so we can design the right product or direct sales promotion or the business plan toward the right customers.

Similar attention can be directed to the rows (customer needs) of Matrix P3. The importance ratings for customer need are summed across all customers. The total represents the total importance across all customers. This total can be normalized. Another measure is to compute the average importance across all customers; this, too, can be normalized to a percent. The matrix proper pre-

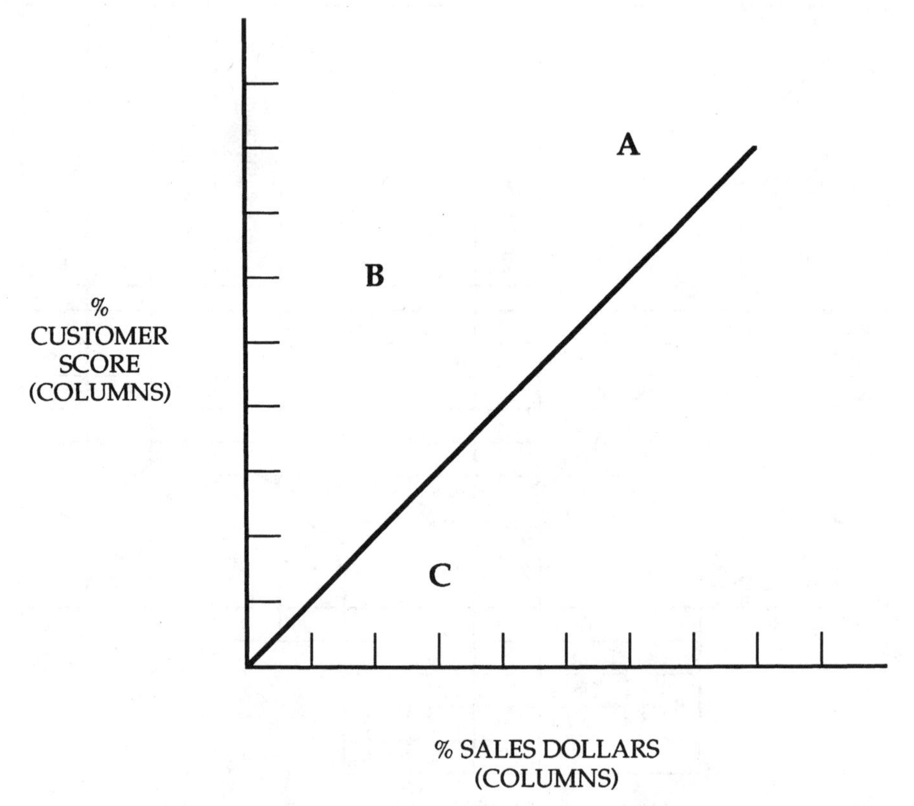

Figure 5.10 Matrix P3 plot, percent customer importance versus percent sales dollars generated from customers.

serves the origins for all averages. It can be used to locate areas of strongest or weakest interaction.

An interesting metric is to post the total budget dollars (usually research dollars) to each customer need. This may take some reaggregating of current accounting figures, but it is well worth the time to do. These dollars should also be normalized.

We can now make another value graph of customer need percent importance and the percent dollars allocated for R&D or other maintenance programs. Figure 5.11 represents such a plot. Are we putting our money in the right place?

PQFD OUTPUT

The output of Matrix P3 is entered directly into the COPC matrix (Figure 5.11) or the HOQ matrix (Figure 5.1). When using the COPC process, the customer needs along with their importance ratings are arrayed with both operational functions and technology. The COPC is then completed as outlined in Chapter 4. The product resulting from a COPC exercise is a product concept(s). These concepts satisfy both customer needs and manufacturing needs. The concept should now be rechecked against the company mission and business plan/participation strategy to be sure nothing has changed (Figure 5.12). If there is a mismatch the concept(s) must be run back through the COPC matrix to see if it can be adjusted so that it fits company mission and plans. If there is a match between plans and concept, the concept is then turned over to the design team and entered into the company total quality management (TQM) system, which consists of many quality tools such as QFD, robust design, design of experiments, failure mode effects analysis (FMEA), and so on, that are used to design, manufacture, and deliver the product. The most popular tool to use next has been FMEA because the COPC matrix vividly illustrates all the important interfaces in the concept. This may also be the appropriate place to subject the COPC outcome (con-

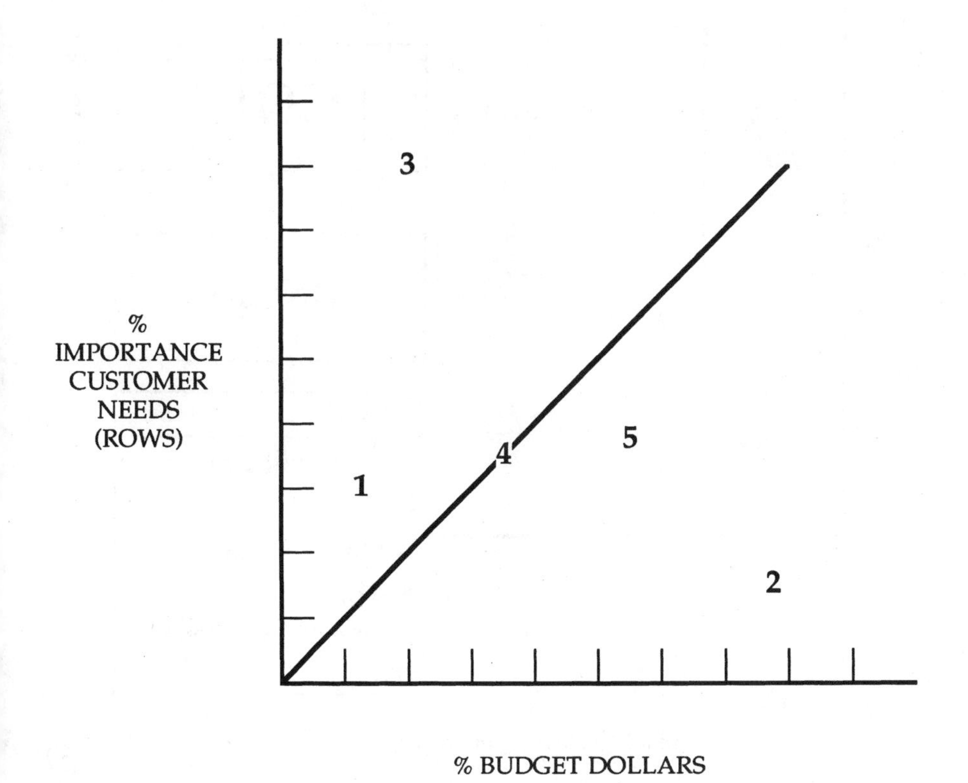

Figure 5.11 Matrix P3 plot, percent importance of customer needs versus percent budget dollars allocated to needs.

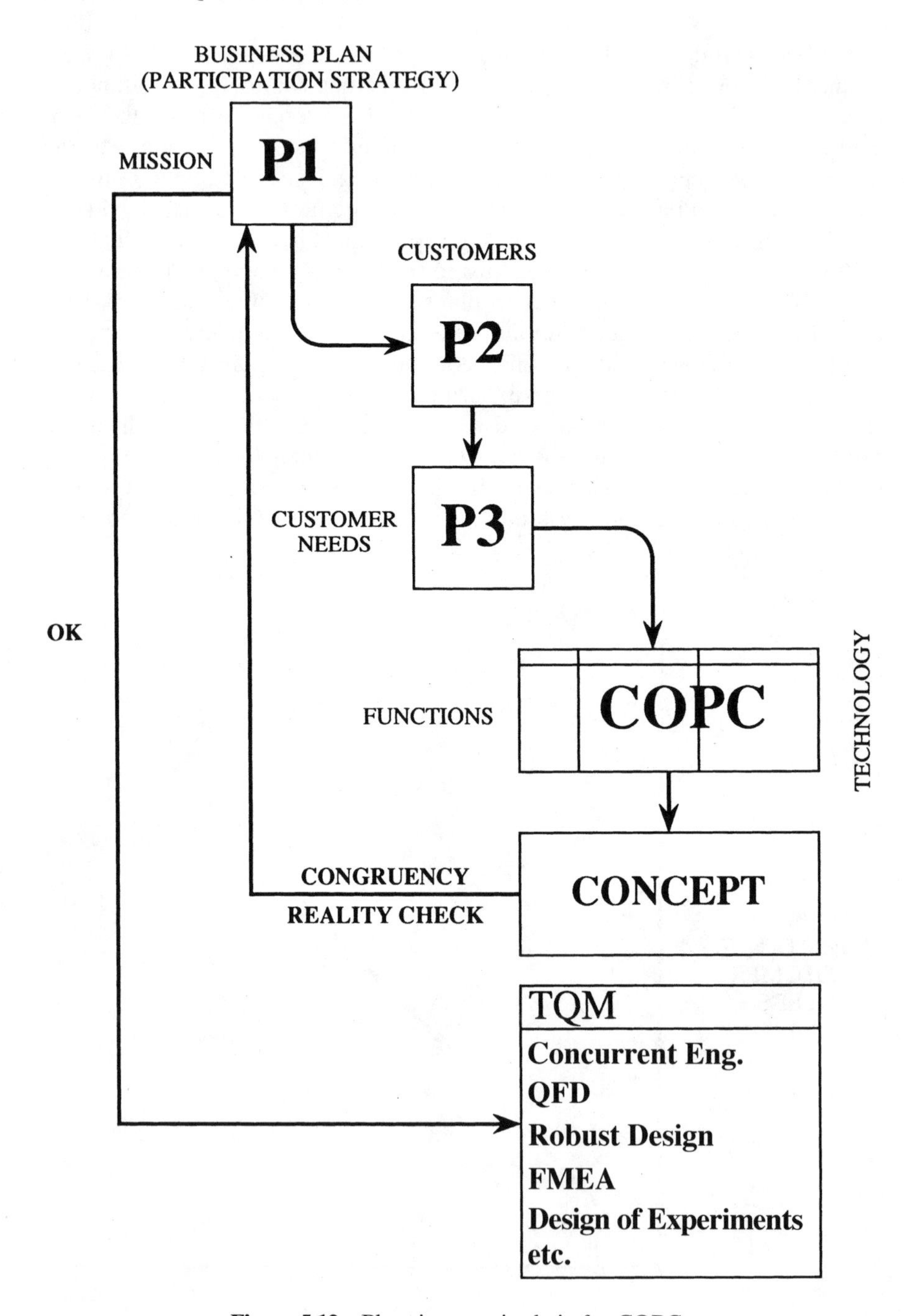

Figure 5.12 Planning matrix chain for COPC.

cept) to decision and risk analysis (DRA) to further build a business case around the concept or particular technical elements of the COPC concept.

SUMMARY

In this chapter I discussed the PQFD planning process. The purposes of this process are: (1) to develop a connection to corporate planning data through a multiple matrix logic trail, and (2) to develop a better customer database to get a better start on developing the VOC. This PQFD process does *not* replace current company planning practices. Rather, the process through structure helps one select and build upon existing corporate planning data. PQFD merely indexes the data down to the HOQ or the COPC matrix to be sure that all data sets are congruent before moving further in the QFD process.

The process through the matrices and the numbers is meant to trigger questions like:

How can this be?
How could the numbers come out this way?
Can we verify that this is so?
Can we justify it?
Is there a logical explanation?
Should we change the rating? How much?
Can we rationalize this discrepancy?

It is these kinds of questions that lead to a sanity check on market and business elements, customer needs, and resource allocation. Most companies today do not have a very good method in place for doing this kind of planning.

The important point is to determine at this up-front stage in the life cycle of a product whether, in fact, we do or do not have a product. Too often "non-products" are left to drag on for too long before they are killed. Consequently, many preventable expenditures are incurred.

Finally, the use of the PQFD process is one way to get management and marketing involved in the process.

BIBLIOGRAPHY

1. Day, R. G., "Using the QFD Concept in Non-Product Related Applications," in *Proceedings, 3rd Symposium on Quality Function Deployment,* Novi, Mich., June 1991, pp. 230–241.

2. Ginder, D. A., “The Strategic Approach to Market Research,” in *Proceedings, 3rd Symposium on Quality Function Deployment,* Novi, Mich., June 1991, pp. 418–428.
3. Lyman, D., Buesinger, R. F., and Keating, J. P., “QFD in Strategic Planning,” *Quality Digest,* Vol. 14, pp. 45–52, May 1994.
4. Olivier, P., “What is the US Total Solution to Product Development?”, in *Proceedings, GOAL/QPC 7th Annual Conference,* December 1990.

CHAPTER 6

TECHNOLOGICAL FORECASTING APPLIED TO QFD

INTENT

In Chapter 3, I expanded the column and row ratings and ratios for several matrices. I even introduced value indices and value graphs to highlight relationships. In Chapter 5, I developed the PQFD model to give us a method to check our matrix development against the business plan and corporate mission. In all of these innovations the mathematical operations were self-contained in the matrix itself. Relationship across matrices was maintained. In fact, the primary purpose of the matrices is to reveal relationships within and across matrices.

In this chapter we discuss tools to examine the influence of QFD/COPC matrix data and signals on external groups, departments, units and companies, and so on. We will specifically borrow several tools from the process of technological forecasting (TF) that developed during the 1970s.

BACKGROUND

"Technological Forecasting (TF) has been defined as a process in which data are gathered and analyzed in order to predict the future characteristics of useful machines, procedures or techniques."[1] Since TF was designed to look into and forecast the future, a large array of modern forecasting methods was developed and came into play. It so happens that many of these techniques can also be used in nonforecasting applications such as QFD. In this respect, TF will be used as a mental model that encompasses visioning, futuring and the examination of paradigms to extend our QFD data into the future.

Some TF tools that are especially useful in QFD are the molecular explosion model, the impact matrix, scenarios, delphi, and monitoring.

MOLECULAR EXPLOSION MODEL

I first learned about this tool in 1973 while attending James R. Bright's Technological Forecasting workshop in Castine, Maine. The tool has had many names, among them impact wheel and futures wheel. Lehning[2] describes and uses her "future wheel" process in the larger context of futuring, that is, looking for consequences and implications. Wagschal[3] also uses the wheel in futuring for exploring consequences of future possibilities. Searle[4] also describes his impact wheel process in detail. Searle uses his impact wheel to distinguish the layers in a hierarchy of consequences and gives explicit instructions in its use. It is very similar to the one about to be described here. Joel Barker[5], the futurist and one of the first to use the tool, has greatly enhanced and refined it into a far more sophisticated technology which is part of his licensed strategic exploration process. I have chosen to use the name "molecular explosion model" (MEM) because of its final finished appearance. The MEM is used to determine the first and higher order impacts on some recipient that can result from an occurrence of some event. Figure 6.1 illustrates a MEM.

The process starts with a center or nucleus that contains the issue at hand. The issue must be stated very precisely. Our example center came from Matrix P1, column "D" in Chapter 5, Figure 5.2. The current column item titled "Convert to new product package and trade dress" was reworded to "Stop new product package and trade dress program." The reason this subject was chosen is that it contributes little or nothing to accomplishing the company mission items as seen in the Matrix P1 relationships. The QFD team now wants to know what will be the consequences to the company if they abandon this project effort.

To develop implications we start with the center and ask: If this event occurs what will be the impact, good or bad, on the company? We record these statements in separate circles around the circumference of the center. Each circle is connected to the center with a single line. Each recorded statement must also be short and specific. A plus (+) or minus (–) sign is recorded with each statement to signify the direction of the impact. Some teams choose to use a scale to indicate both direction and magnitude of the impact such as ±1, ±3, ±5. See Figure 6.2. The correct way to show impact is the method that works best for the team. There is no inherently right or wrong way.

Now that the team has completed level one, they repeat the process all over again on each of the new Level 1 circles. This time the new (Level 2) circles are connected to the Level 1 circles with a double line. Finally, a third level impact is derived by expanding all of the Level 2 circles, and so on. Three lines connect Level 2 circles to Level 3 circles. Many times several Level 3 impacts are expanded further, as seen in Figure 6.1. Recording of the impact statements must be done in a circular manner as opposed to a linear chain of events. That is, expand MEM events in concentric circles. It is easy to fall into the trap of recording a Level 1 event and then continuing to expand this event on out to a forth or fifth level before going to the next Level 1 event. Most teams will have exhausted their thoughts by the end of round 4. Depending on the subject and use, many times two levels are sufficient. The amount of detail and number of levels is determined by the subject and needs of the team.

How do we use this information? What we are looking for are events or a chain of events that can have a serious positive or negative impact on the company, something for which no one was prepared. We are looking for serious

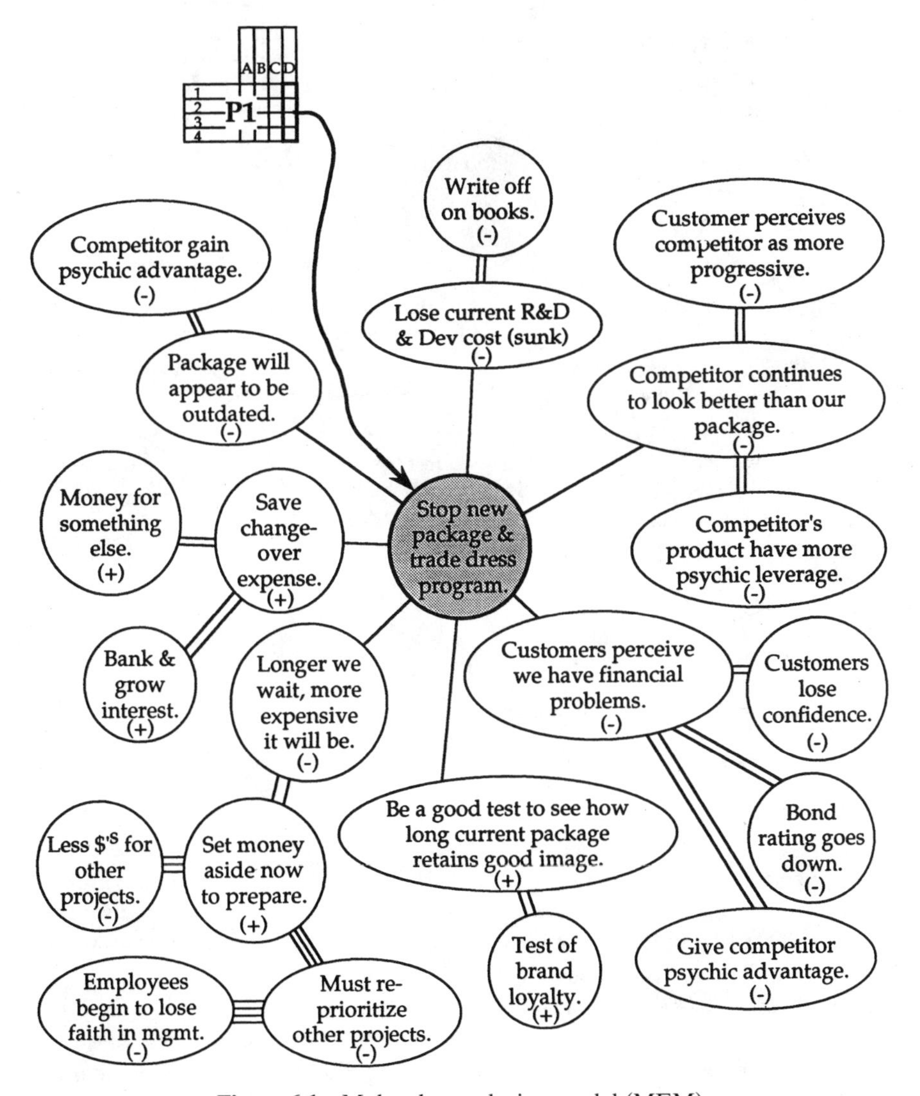

Figure 6.1 Molecular explosion model (MEM).

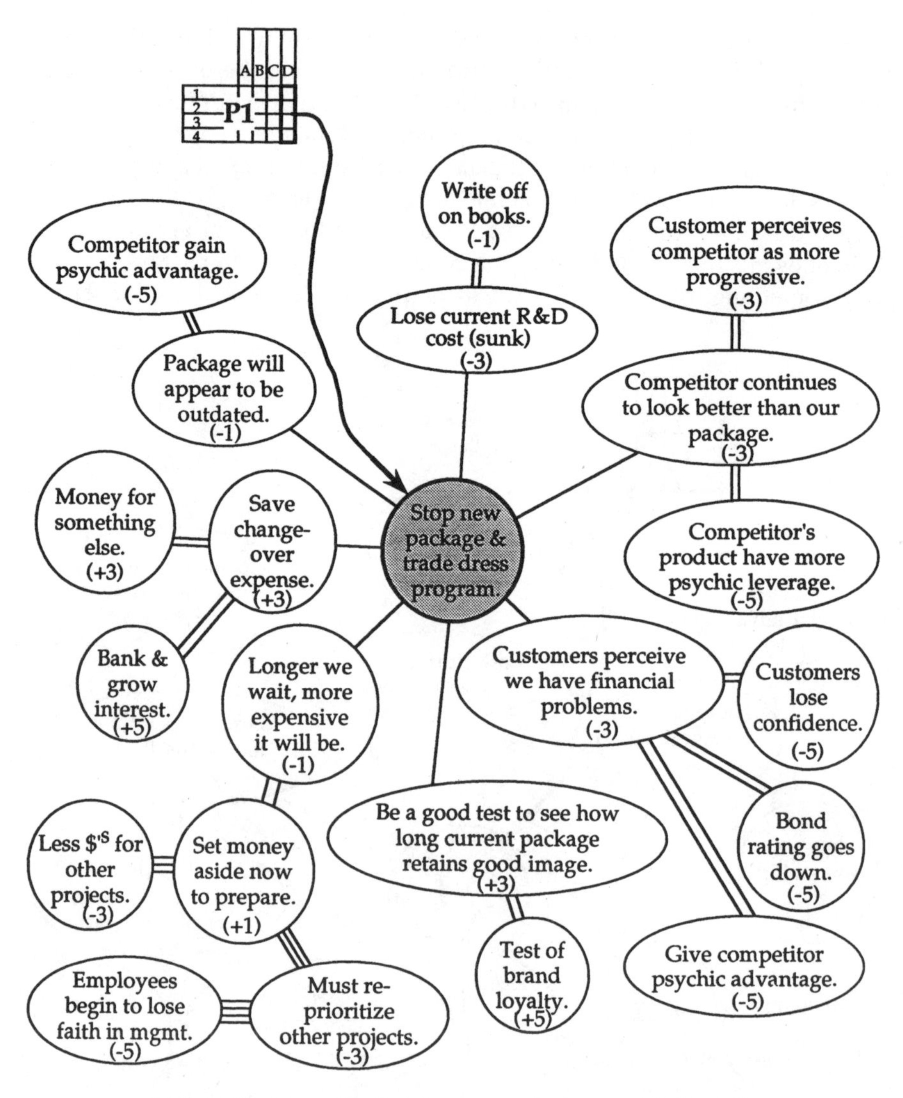

Figure 6.2 Quantified molecular explosion model (MEM).

consequences that may result from some action or event. Many of these events are surfaced during a QFD project. The MEM then serves as a testing ground to determine the impacts of the events before they occur. The many QFD matrices described in this book generate and contain a tremendous amount of information. In essence, we are dealing with information overload. Many signals are discovered because of the systematic procedure for deriving the QFD matrices. When an important signal is discovered during the process of developing the matrices, it should immediately be highlighted with a yellow marker so it is not lost later on as more data are developed. When the matrix/matrices are finished, the team can return to those marked items and apply the MEM to envision the ramifications of those signals on the company, business unit, and so on. Once the implications are structured the team can then develop and recommend appropriate responses to those signals.

Other creative uses of MEM are to leave the chart on the wall so other members and nonteam members can observe it and add to it; sort of a "what's missing in the picture" exercise. Another useful exercise is to convert the wheel to prose/paragraph form and send to people over an electronic mail system and ask for responses.

I have chosen to illustrate the MEM in conjunction with a P1 matrix. It can just as well be used with all the other matrices. Give it a try.

IMPACT MATRIX

Another useful TF tool is impact analysis that incorporates an impact matrix. Impact analysis is a subjective quantification technique for evaluating the impact of external factors on a set of items. It is a matrix scoring method wherein raters use (+)'s and (−)'s and zeros to quantify their perceptions of interactions of items. It can be used in any situation where it is desirable to measure both the positive and negative forces that impinge on a system or set of alternatives.[6,7] An impact matrix, then, is used to express the impact of the rows on the columns (Figure 6.3).

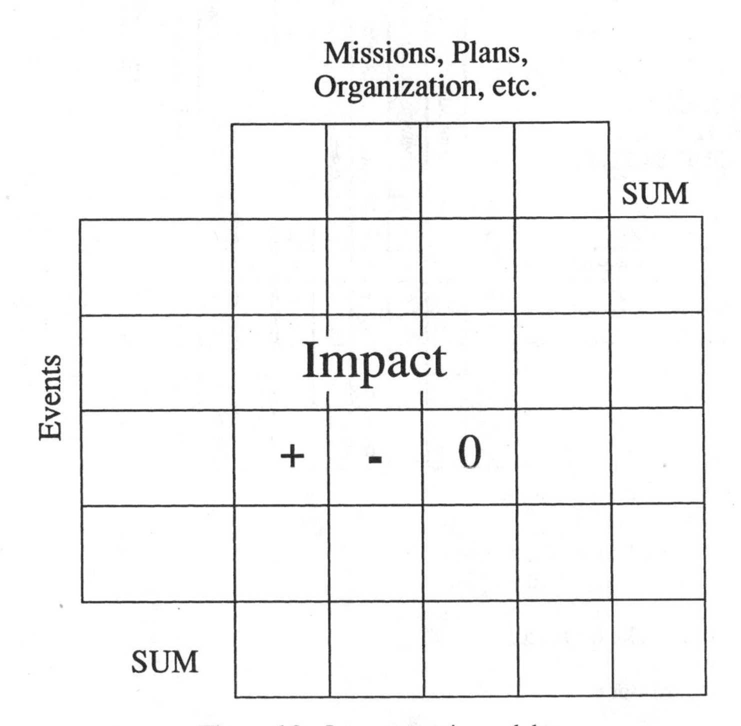

Figure 6.3 Impact matrix model.

How might this be used in QFD? It can be used in Matrix 1 (HOQ) to determine the impact of certain selected product technical requirements (PTRs) on the corporate mission, business plan, marketing plan, and so on. Usually the higher scoring PTRs would be selected to become the rows of an impact matrix.

Let us walk through the example in Figure 6.4. PTRs from Matrix 1, Chapter 2, Figure 2.1, were selected so as to determine the effect of the PTRs on the corporate mission and business plan, which were, in turn, taken from PQFD Matrix P1, Chapter 5, Figure 5.2.

PLANNING / PRODUCT TECHNICAL REQUIREMENTS		MISSION					BUS PLAN						
		Create Superior Product Val.	Leverage Core Platform	Focus on Customer Satisfaction	Achieve Mkt. Leadership	Enter, Grow New Business	Eliminate ABC Defects	Reduce RT to 2 Hours	Open 4 Locations in Territory XYZ	Convert to New Product Package & Trade Dress			
		A	B	C	D	E	F	G	H	I	Σ+	Σ-	Σo
1	Lead Dust Generated	+	o	+	o	o	o	o	o	o	2	0	7
2	Time Between Sharpening	-	o	-	-	o	o	o	o	o	0	3	6
3	etc.	+	o	+	+	o	o	o	o	o	3	0	6
4	etc.	+	+	+	+	o	o	o	o	o	4	0	5
	Σ+	3	1	3	2	0	0	0	0	0	9		
	Σ-	1	0	1	1	0	0	0	0	0		3	
	Σo	0	3	0	1	4	4	4	4	4			24

Rating Scheme

(+) = Will make it easier to achieve mission and business plan item.

(-) = Will make it difficult...

(o) = No influence.

Figure 6.4 Impact matrix example.

To determine the impacts, we begin with row one, "Lead dust generated," and ask: will "Lead dust generated" have a positive, negative, or no effect on "Create superior product value"? In this case we feel it will have a positive effect. That is, improving or changing the PTR as indicated on the HOQ will make it easier for us to accomplish the specific mission item. Conversely, a negative impact would make it more difficult to accomplish our mission item. Likewise, "Lead dust generated" has no impact or influence on accomplishing our second mission item, "Leverage core platform."

The entire matrix would be quantified in the same way. It is best to complete the matrix by scoring one complete row at a time until finished. Once quantified, the matrix columns and rows are summed to determine the total (+)'s, (–)'s, and zeros. The column totals show the combined effect of all of the PTRs on each of the mission and business plan items. When questioned on totals, there is an audit trail back into the matrix to determine where the totals came from. Which PTRs had the most influence? Also, row totals show the combined overall affect of PTRs across all mission and business plan items.

What do you look for with the information derived from this mechanism? In the matrix (Figure 6.4) we can see that the PTR "Time between sharpenings" has negative impacts on three out of five mission items. This is a signal that more discussion is required. Upper management should be consulted. Should we continue funding this PTR? How much, if any, should we change the parameters affecting lead dust generated? Why are there negatives? How could this be? Is there a mistake? Let's aim before we shoot. Notice that none of the PTRs has any influence on the business plan items. Why? What happened? It could be that the PTRs fully support and are in alignment with the business plan items. They just won't make it easier or more difficult to achieve them.

What are some other applications of the impact matrix? It could be used again with the PTR columns by estimating the impact of PTR target values, or the impact of other manufacturers' PTR measures on our sales or business plan. For example, what if several other manufacturers' measures are much better than ours? If we do not close the gap, what impact will that have on our mission and business plan? This is also an excellent place to use a MEM discussed earlier.

There are several ways to quantify an impact matrix. First of all, instead of just simple (+) and (–), one could use numbers from an impact scale such as ±9, ±3, ±1, and zero. The PTRs could be entered along with their respective HOQ scores. The normalized PTR scores would then be used as weighting factors to multiply against the impact scores to derive a weighted impact.

MONITORING

Monitoring is a TF surveillance technique that entails keeping current tabs on significant signals in order to assess change.[8] The HOQ and the COPC are excellent places for discovering signals to watch for future significant changes.

Likewise, a monitoring program can be started from an impact matrix or a MEM, which themselves could also have been developed from a QFD or PQFD matrix.

The basic need is to establish a file, or an information storage and retrieval system, and to assign responsibility for maintenance of the file. Using monitoring, QFD signals can be followed for changes. The most obvious type of monitoring is: Are breakthroughs occurring in any of the new technologies?

Data are collected from the general literature, as well as annual reports, patents, and personal observations and contacts. Consideration should be given to data storage and retrieval, as simple notebooks and files may quickly become impractical.

A monitoring database, depending on the monitoring subject, really forms the basis for marketing and/or competitive intelligence. If a competitive or market intelligence department already exists in the company, then the QFD team should share their signals and uncertainties with them and have them do their own equivalent of monitoring. Don't reinvent the wheel. The point is that the use of QFD/COPC and TF can play an important role in keeping the company alert to changes and emerging trends that could have a significant impact on the business. QFD is the process that uncovers the uncertainties to monitor.

DELPHI

Delphi is a TF inquiry tool for looking into the future. It is used to ask people what events they see happening in the future, their likelihood of occurring, and their impact if they do occur. It is an excellent tool for studying the trends and changes in things like customer needs (voice of the customer—VOC) and technology. Delphi can be used to extend signals into the future that have surfaced through the MEM or the impact matrix applications. This method is covered in more detail with an example in Chapter 9.

SCENARIOS

One way of integrating the signals and data derived from the application of the various TF techniques to QFD is to write a series of scenarios. Scenario writing is a qualitative forecasting technique that is used to project various futures from present conditions based on stated assumptions. Data from the MEM, impact matrix, monitoring, and Delphi techniques provide excellent grist for writing scenarios.

Scenarios may be written in many different forms, lengths, and formats. Generally a set of several alternate scenarios is constructed as opposed to one single most-likely scenario. The three most used styles are a pessimistic scenario, an optimistic scenario, and a most-likely scenario. They may be written in prose form or in a matrix format (Figure 6.5). Regardless of format they should be believable, relevant, and thought-provoking. Hart[9] notes that scenarios can be a powerful means of forcing people to look at alternative futures because scenarios presume events, especially unpleasant ones, have occurred. Scenarios can be a useful method to observe the future of VOC data.

Basic input for writing scenarios may be taken from the MEM model discussed earlier or from the impact matrix.

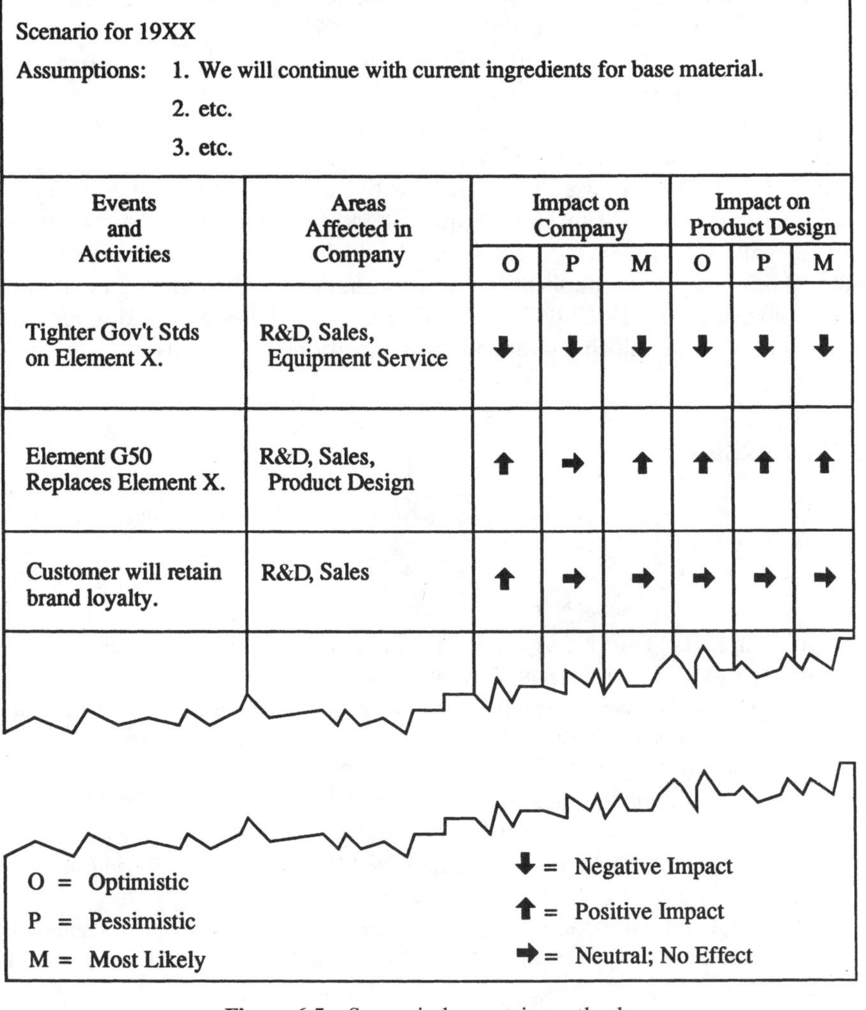

Scenario for 19XX

Assumptions:
1. We will continue with current ingredients for base material.
2. etc.
3. etc.

Events and Activities	Areas Affected in Company	Impact on Company			Impact on Product Design		
		O	P	M	O	P	M
Tighter Gov't Stds on Element X.	R&D, Sales, Equipment Service	↓	↓	↓	↓	↓	↓
Element G50 Replaces Element X.	R&D, Sales, Product Design	↑	→	↑	↑	↑	↑
Customer will retain brand loyalty.	R&D, Sales	↑	→	→	→	→	→

O = Optimistic
P = Pessimistic
M = Most Likely

↓ = Negative Impact
↑ = Positive Impact
→ = Neutral; No Effect

Figure 6.5 Scenario by matrix method.

Another interesting exercise using scenarios is in conjunction with the HOQ, particularly the rows and the planning matrix. For example, the team, with help from marketing, could use the scenario technique to write future advertising brochures and literature based on competitive analysis, improvement ratios, customer importance, and sales point. The same sales approach could also be used nicely with the COPC matrix, especially the various technology paths that are developed. Each different technology path could be a separate scenario.

SUMMARY

In this chapter I have shown how some of the tools from TF can be used with QFD. These tools are used to determine the effect of specific QFD items on external events and organizations. The TF tools are not used for forecasting. They are used for creative thinking about the future, which can help us document possible effects of decisions made from QFD data. When signals from the value graphs are discovered, their effect can be tested using the TF techniques. That is, TF and VE are used to locate sensitive areas and to evaluate them for outcomes and impacts. The TF tools are not and should not be used in a predictive capacity but rather in a thought-provoking what-if mode. Their purpose is to help us think about the implications and consequences of what may be implied from some of the QFD, COPC, and PQFD matrices. They are used to create a mental model for exploring interactions, implications, and trends.

REFERENCES

1. Martino, J. R., *Technological Forecasting for Decision Making,* Elsevier North Holland, New York, 1972.
2. Lehning, M. A., "Managing Change," Tracking Trends, Inc., Suite 1900, First National Center, Omaha, Nebraska, 68102.
3. Wagschal, P. H., "Futuring: A Process for Exploring Detailed Alternatives," *World Future Society Bulletin,* September–October, 1981, pp. 25–32.
4. Searle, B., "The Impact Wheel: the Empowerment Experience," *The 1989 Annual: Developing Human Resources,* University Associates, University Associates, Inc., 8517 Production Ave., San Diego, Calif., 92121, pp. 83–87.
5. Barker, J. A., Infinity Limited, Inc., 8311 Windbreak Trail, Lake Elmo, MN 55042-9521. Author's conversations with Mr. Barker.
6. Shillito, M. L., and DeMarle, D. J., *Value: Its Measurement, Design, and Management,* John Wiley & Sons, New York, 1992.
7. Shillito, M. L., "Impact Analysis," paper presented at James R. Bright's Technology Forecasting Workshop, Castine, Maine, Industrial Management Center, Austin, Texas, June 1977, p. 36.

8. DeMarle, D. J., and Shillito, M. L., "Technological Forecasting," in G. Salvendy, Editor, *Handbook of Industrial Engineering,* John Wiley & Sons, New York, 1992, Chapter 11.1, pp. 11.1.1–11.1.15.
9. Hart, J. L., "Technological Forecasting, From Board Room to Drawing Board," *Machine Design* **48** (3), 90–93, February 12, 1976.

BIBLIOGRAPHY

1. Porter, A. L., Roper, A. T., Mason, T. W., Rossini, F. A., and Banks, J., *Forecasting and Management of Technology,* John Wiley & Sons, New York, 1991.
2. Vanston, J. H., Jr., *Technology Forecasting: An Aid To Effective Technology Management,* Technology Futures, Inc., Austin, Tex., 1982.

CHAPTER 7

ORGANIZING AND LAUNCHING A QFD PROJECT

FORMAT OF THIS CHAPTER

This chapter is to serve as a checklist for practitioners to use in planning and launching a QFD project. The contents are discussed in sequential order of occurrence for planning the project. This chapter is to be an operational document and guide. Chapter 8 is a backup reference document that discusses these subjects, plus others, in considerable detail.

BACKGROUND

Organizing and launching a QFD project properly is critical to its success. A poor job in preparation can jeopardize the outcome and success. The point to be made is that you cannot just cold-start the QFD team by immediately building the house of quality (HOQ). You can't just start by brainstorming customer needs. You can't just put a team of people together and expect them to understand or agree on content and work smoothly with each other. You cannot have a poorly stated or nonexistent team mission and scope and expect the team to magically discover a focus on their own. Unstated or undocumented assumptions will continuously stall or derail team progress and momentum. And finally, you can't put a new group together and expect everyone to initially agree on the product to be studied, let alone the market, segment, user, and chief buying influence. The QFD facilitator must consider all of these items and work well ahead of the QFD project launch date with the project or team leader, or both, to plan the project.

Let us now establish what a QFD consultant must do to ensure success of QFD projects and applications. Figure 7.1 is a flow chart of events and activities that outline a sequential order of events necessary for a successful QFD project launch. This figure forms the outline for this chapter.

QFD activity is initiated because of some need. This need finally reaches you through some form of request. You, the QFD consultant, schedule a meeting with the requestor. At this meeting it is important to clarify the objective or purpose for using the QFD process, because too often the requestor

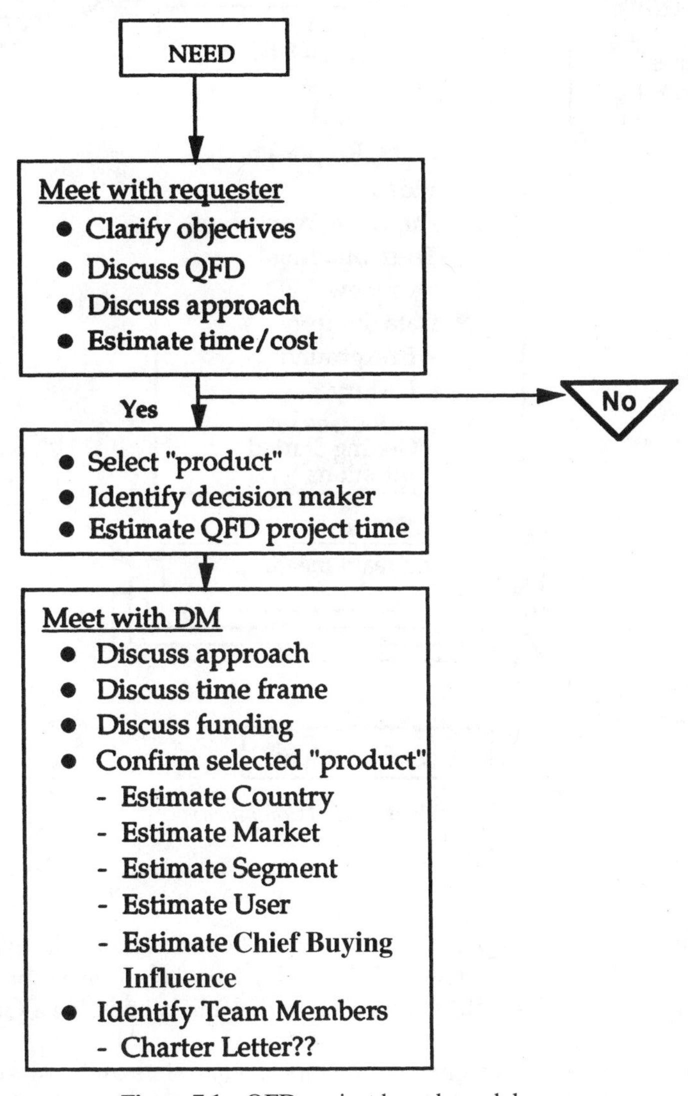

Figure 7.1 QFD project launch model.

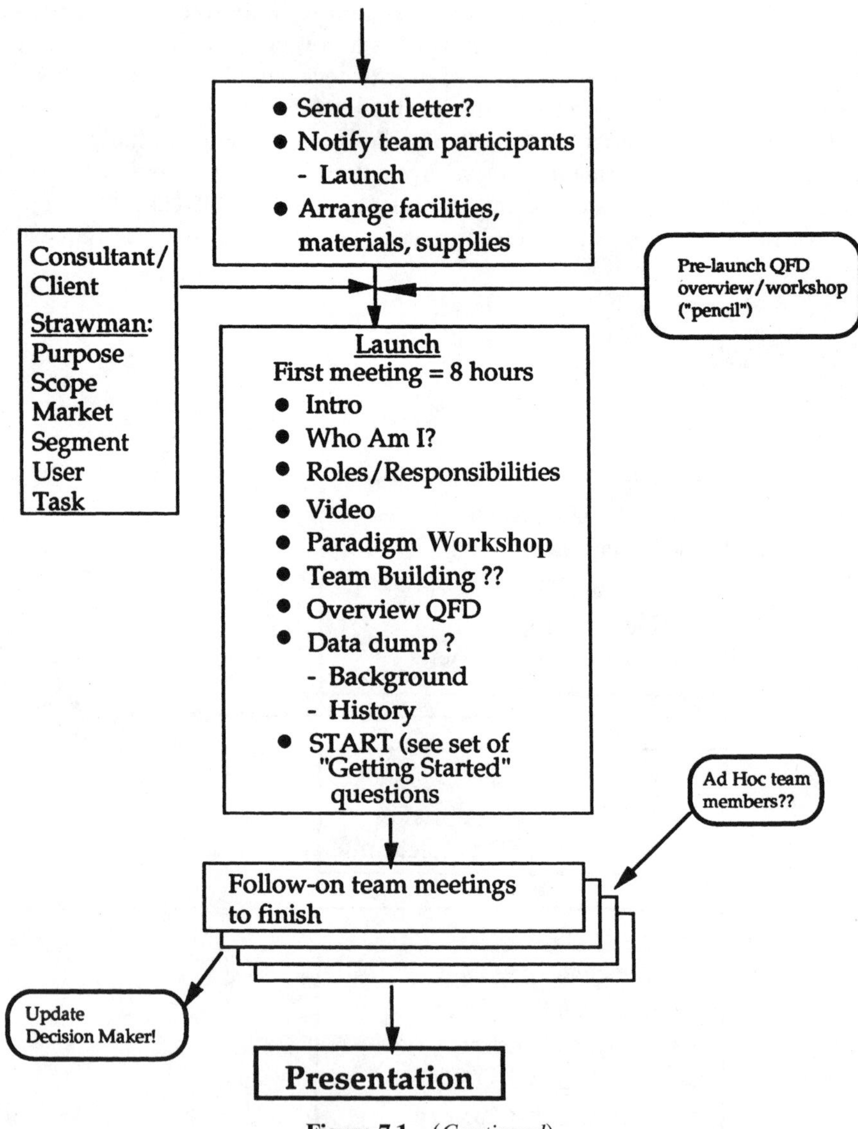

Figure 7.1 (*Continued*)

knows little (or nothing) about QFD. It may be necessary at this point to give a brief (20 minute) overview of QFD with particular emphasis on the HOQ. Next the approach and mechanics of initiating and conducting a QFD project are outlined.

This is followed by estimating the approximate time and cost to conduct a QFD study. At this point the requestor may elect not to do the QFD project.

If the requestor is not too scared and survives, another meeting is scheduled to discuss preparation in more detail. The product that is to be the subject of the QFD study is defined thoroughly, along with the country, market, segment and customer. It is normal for many unanswered questions to appear at this time. Fortunately these unanswered questions reemphasize the need for thorough preparation for the study. Figure 7.2 is an outline of 12 "getting started" questions needed to initiate a QFD study. These questions are reviewed with the requestor to develop a "strawman" document of the best answers available at this time. These questions will quickly establish what is known and what is not known about the subject of the study at this time. They will also make it

1. Purpose:
   (a) Why are we doing this study?
   (b) What is the team mission?
2. Completion date:
   (a) When *must* this study be finished?
3. Decision maker:
   (a) Who is the decision maker?
   (b) Who is the first person in the decision chain that can say "no"?
4. Scope:
   (a) What is included in this study?
   (b) What is not included in this study?
5. Product:
   (a) What product? Model? Generation?
   (b) World class? Revolutionary?
6. Market/customer:
   (a) Who is the customer we are trying to satisfy?
   (b) Country?
   (c) Market?
   (d) Segment?
   (e) User?
   (f) Chief buying influence?
7. Time horizon for the product?
   (a) This year? Next year? When?
8. Assumptions:
   (a) Product
   (b) Market
   (c) Company
   (d) Manufacturing: Who? Location?
   (e) Distribution
   (f) Customer
   (g) Other

Figure 7.2 QFD starting questions. These questions are used to initiate a QFD project. Discussing and documenting these subjects is one of the most important parts of the QFD process. A poor job here can cause teams to be off-course, lose time, and develop excellent recommendations on the wrong thing.

9. Organization business plan:
   (a) Do the answers to the questions above fit the organization's business plan?
   (b) Do we (the team) have a copy of the business plan?
   (c) Are there spinoffs that will apply to other organizations?
10. Team members:
   (a) Based on the answers to the questions above, do we still have the right people on the core team?
   (b) The core team remains for the life of the project; who are the members? What background do we need? What geography represented?
   (c) Ad hoc members? Who? What? Background/expertise/information? When do we need them? How long do we need them for?
11. Buying/purchasing drivers (influences):
   (a) Economics
   (b) Product performance
   (c) Safety
   (d) Ease of use
   (e) Workforce capability
   (f) Environment
12. Task/deliverable/function:
   (a) What task is the customer/user trying to accomplish through buying our product?
   (b) What is the deliverable or output from using this product?
   (c) What is the basic function of the product? Why does it exist?

Figure 7.2 (*Continued*)

necessary to determine the scope of the study; that is, what is included and what is not included in the study? Based on the additional information, a better estimate of time, number of people, and cost is determined. In addition to the number of people for the team, actual names, backgrounds, skill sets, and geographic areas are considered.

Next the decision maker is identified. To whom will the future QFD team present its recommendations? Who will be the first person in the decision chain to approve or disapprove?

It is now time to meet with that decision maker. The QFD facilitator and the requestor put together a recommendation to proceed with the QFD study. The strawman document is based on the answers to the 12 questions. The starting questions and their relationship with the decision maker and business are illustrated in Figure 7.3.

MEETING WITH THE DECISION MAKER

The requestor, with backup support from the QFD facilitator, meets with the decision maker. This meeting is used to discuss the QFD approach, time frame, cost, and the product related information. The market and customer related

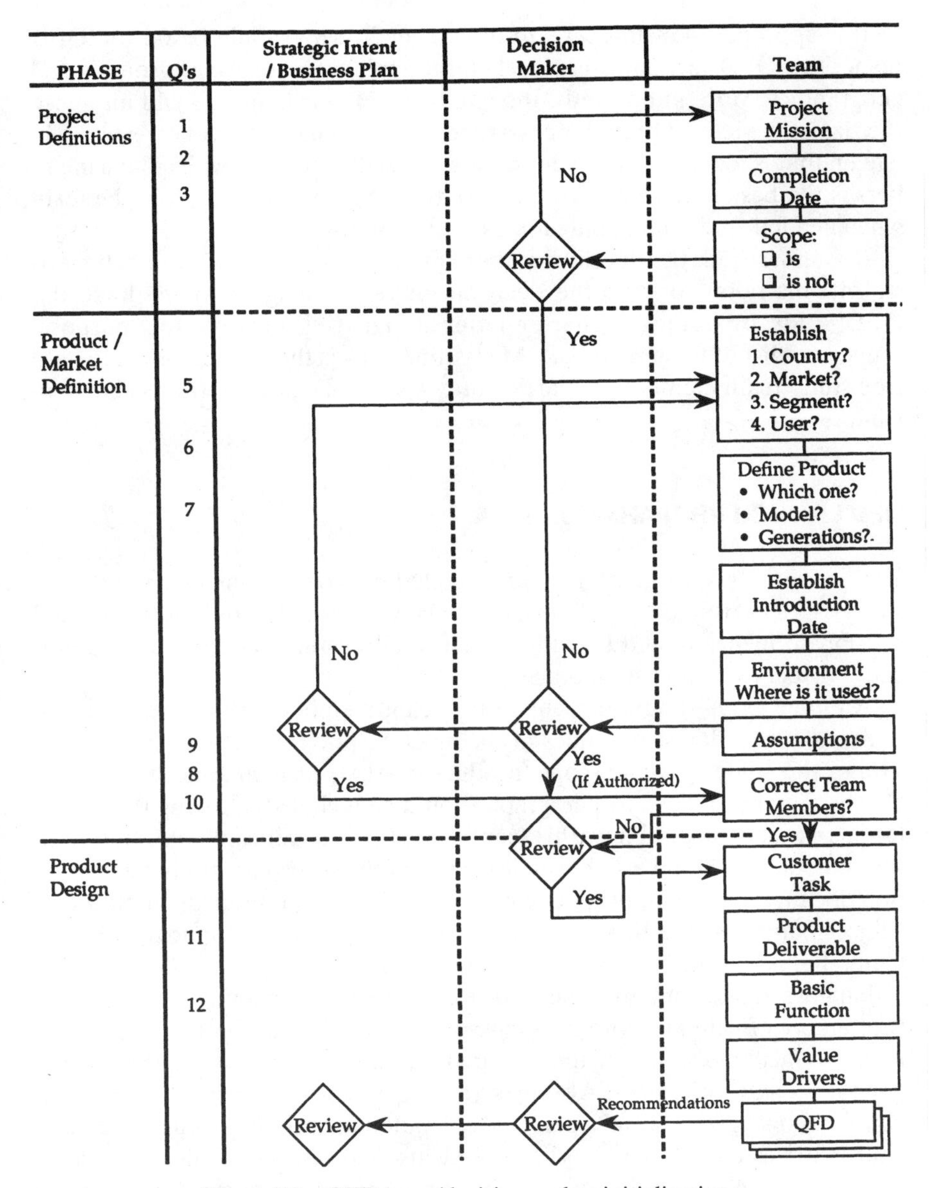

Figure 7.3 QFD team/decision maker initialization.

questions are especially important to be sure we will be working on the right thing. Many course corrections are given at this point. There are many embarrassing moments as well, especially if it is discovered that the decision maker does not have many of the answers. This, however, is the beauty of the process. We discover what we don't know and join forces in a collaborative structured process to seek the answers.

It is appropriate at this time to discuss again who should be on the QFD team. The QFD facilitator can be helpful here, suggesting backgrounds and geographics. More importantly, though, the QFD facilitator should influence the size of the team. Often there is a tendency to include everyone, resulting in a team that is too large to make progress. A small core team with ad hoc members is the best compromise (see Figure 8.4). Ad hoc members join the team when needed, and remain only for a specific duration.

If all goes well, the decision maker agrees to proceed. Often it is beneficial at this point to have the decision maker (management) endorse the QFD study by issuing a charter letter designating the sponsor, purpose, scope, and list of team members. Many times this is the needed boost to help free study team members to participate. A sample charter letter is shown in Chapter 8.

QFD LAUNCH PREPARATIONS

The decision maker's charter letter is mailed out to team members and their supervisors as well as the other individuals who should know and who could possibly influence the QFD study. Yes! The QFD study should also be politically correct to minimize sabotage.

Another letter should be sent to the team members listing the locations, building, room, dates, and times of the meetings(s). The latest strawman version of the study purpose, scope, product, market, segment, and user should also be appended. A short paragraph should be included outlining the expectations of the team and how this team can make a significant contribution to the company. It should also be made clear that the strawman product information document is a temporary document to get the team thinking about those subjects, and that the final document must developed and approved by the team; that is, it must be their collaborative information.

Finally arrangements must be made for facilities and materials. A large room (preferably off-site and with adequate parking) with lots of wall space is preferred to facilitate a QFD team. Be sure to bring a good supply of chart paper, pens, tape, and sticky pads. Also make arrangements for a video tape player if you are showing videos and a personal computer and printer if you are building your matrices on-line. An overhead projector and screen will also be used to present data. Additional blank overhead transparency material and pens are also used frequently. Finally an in-room pickup breakfast should be arranged, along with refills all day and munchies in the afternoon.

THE LAUNCH

The launch meeting is generally scheduled for eight hours. It often begins with a kickoff by upper management stressing the importance of the project to the company and giving additional background information. At this point the QFD facilitator begins to run the meeting. Introduction of team members is accomplished by a "who am I?" exercise discussed in Chapter 8. Roles and responsibilities are established. The QFD facilitator is the process consultant and keeps the team on track. The facilitator is not a content expert, however, and does not provide all the energy for the project. The team members, with facilitator assistance, must establish their roles as well. The video by Joel Barker, *Discovering the Future: The Business of Paradigms*[1] is shown, after which there is a paradigm workshop. The purpose of this workshop is to surface existing paradigms that can affect the way the team perceives customer, product, and corporate operational and business policies. The objective is to make visible those paradigms that make it difficult for us to look into the future with an open mind to design products or service. This list of paradigms is posted at all of the follow-on meetings. Keeping them visible makes team members conscious about drifting back into the paradigms and applying them at the meeting.

If team members are not familiar with each other, sometimes it is necessary to conduct some team building exercises as noted in Chapter 8. An overview of QFD is given for the team. Chances are that several members may not know about QFD or how it works. There are two options for teaching the team about QFD. As mentioned above, a half hour to 45 minute overview can be given at the launch meeting. Another option that has worked very well is to conduct, for the QFD team, a half day seminar on QFD prior to the launch. This seminar consists of an overview plus a three hour workshop where the team actually builds a small HOQ. We use a pencil as the product and start by brainstorming customer needs, then construct an affinity diagram, and continue to finish the matrix. The size of the matrix is restricted to 5×5. At the end of the session, usually conducted the afternoon before the QFD launch, the team is versed in the HOQ and is partially up to speed regarding process. They then come to the QFD project launch well versed on what they have to do. My experience shows that this prelaunch QFD workshop considerably accelerates the pace of the team and shortens team project cycle by about two days of team time.

After the QFD overview the facilitator conducts what I call a "data dump." Marketing, manufacturing, research, sales, design, and other pertinent areas are invited to present the latest data directly related to the project/product under study. It is important that team members are aware of and have access to this data. Generally these presenters give the team copies of their presentation. The team has to be up on the latest on everything that can affect the project and their decisions.

Probably most important of all, the director of the business unit is also asked to participate in this data session by sharing the latest version of the business plan. The team must know whether their thinking and possible recommendations fit the business plan. Too often teams are not privy to this information, as we see later. The team, usually through the team leader, will update management on its progress and direction. Early on it is important to check the team's emerging design or recommendations against the business plan for congruence. Is there deviation or a disconnect?

Once the data dump session is completed, it is a good idea to take a break. Sometimes the data dump can be very voluminous and team members will need a break. After the break is when the new team starts to get into the heart of the QFD project. They begin by addressing the strawman document with the answers to the 12 questions (see Figure 7.2). Each question is debated in detail until there is team agreement. The final team-developed 12 questions document then becomes the foundation for beginning the QFD HOQ. However, even at this stage there may not be agreement on all questions, and, as a result, many action items may be assigned to team members for seeking further data for clarification. The team, though, will proceed to build the HOQ based on the current assumptions that are slated for clarification and updating. The important thing is that the gap in information or missing answers has been properly surfaced for attention and action. Many times, without QFD, these information gaps remain unnoticed until well into a project; then they surface, usually at the worst possible time.

At this point, after this initial meeting, it is important to review the answers to the 12 questions with the decision maker in order to get any course corrections early in the project. The review is usually done by the team leader.

POST LAUNCH FOLLOW-UP MEETING

It is highly likely that the entire launch agenda may not be completed in one day. Follow-up meetings are scheduled for the next six to eight weeks. The follow-up meetings are four hours in length and are held at least once per week, preferably twice a week. These meetings should be set up on a regular schedule so everyone knows the meetings all occur on the same day at the same time. This makes it less confusing when arranging calendars for other business.

At these follow-up meetings ad hoc members who have specialized information will be called in to supplement the core team and will then fade out when they are finished. The team leader must regularly update the decision maker on the follow-up meeting, progress, and direction so early course corrections can be made if necessary. It is during these follow-up meetings that the HOQ is built as well as any of the other three matrices. As assumptions are confirmed these data are entered into the matrices.

POST HOQ VERIFICATION

When the HOQ is finished all details are checked. Pending questions are reviewed; action items are checked to be sure all have been completed. There may be some items in the HOQ that need further investigation. It is at this time that focus groups or marketing surveys are scheduled, based on data and unknowns that have surfaced during construction of the HOQ. Marketing schedules these surveys. After they are conducted the data are reviewed with the QFD team. The data are used to verify the data in the HOQ. New items that surfaced on the marketing interviews or focus groups are also entered into the HOQ. The team now schedules another HOQ review meeting with management to review the updated HOQ as well as the focus group or survey data. This meeting very often reveals conflict with decisions as well as unknown and sometimes disturbing information. The facilitator plays a special role in minimizing conflict and arbitrating dissension. Just as often the team and management discover that they have done a good job and that the focus group information verified their HOQ data and assumptions. Again, this is the beauty of the process; it tells you what you know and what you don't know. It tells you what to verify and what additional data to collect. The verification session with management is critical because it will force verification, decisions, and course corrections.

When management and the team feel comfortable with the updated HOQ, the team now begins to construct Matrix 2. The core team will remain intact. New ad hoc members may be added for Matrix 2. Since Matrix 2 involves product subsystems, parts and/or components, the new ad hoc members are usually from the product design and prototyping community.

Likewise, the team proceeds to build Matrices 3 and 4. Verification with the decision maker is sought through all matrices. Ad hoc members will be exchanged at appropriate places throughout the QFD project.

SUMMARY

In this chapter I have described how to successfully launch a QFD project. I have outlined one approach that I have used with considerable success. I cannot overstate the importance of a properly prepared project launch. A poor job preparing and conducting the launch can cause teams to be off-course, lose time, and develop excellent recommendations on the wrong thing. The periodic involvement of management and the decision maker is also critical to the success of the project. The team leader, with help from the QFD facilitator, can be proactive in nurturing management involvement, which in the end assures project success and implementation of the team's recommendations. Management reviews throughout the QFD project foster the interest and participation of management, which in turn engenders ownership and buy-in.

REFERENCES

1. Barker, J. A., *Discovering the Future: The Business of Paradigms,* Video, Charthouse Learning Corp., 221 River Ridge Circle, Burnesville, Minn. 55337, 1984.

CHAPTER 8

QUALITY FUNCTION DEPLOYMENT: BEHAVIORAL AND ORGANIZATIONAL ASPECTS

BACKGROUND

Everyone today is involved one way or another in change. In fact, one constant we can always expect in the future is change. Because we have change, we will always require decision making. Decision makers very often deal with complex issues. All decision makers can always use more help in dealing with issues and generating and evaluating alternatives. Generally, there is a reluctance to deal with complexity in a structured systematic manner. According to the John Warfield model,[1] there are three basic elements to addressing an issue: the issue itself, an interdisciplinary team, and an appropriate methodology (see Figure 8.1).

As in any triangular relationship, one element cannot work well without the others. They are all interdependent. The connections between the elements are just as important as the elements themselves. If a flexible simplistic methodology can be developed that can serve both the team and the issues, then there is a greater chance of connecting the team successfully with the issue. Fortunately, QFD provides this connection. The more multidisciplinary the QFD team, the better the connection. However, the interconnections of the three elements involve people, teams, organizations, and politics and thus further compound the relationships (Figure 8.2).

The successful application of QFD projects and methodology is very often weakened because the behavioral, people, political, and organizational elements are not addressed properly. This chapter discusses some of these issues that affect a QFD project and a QFD program.

Some of the many questions that should be considered in launching and conducting QFD projects are: What is the scope of the study? Define what the

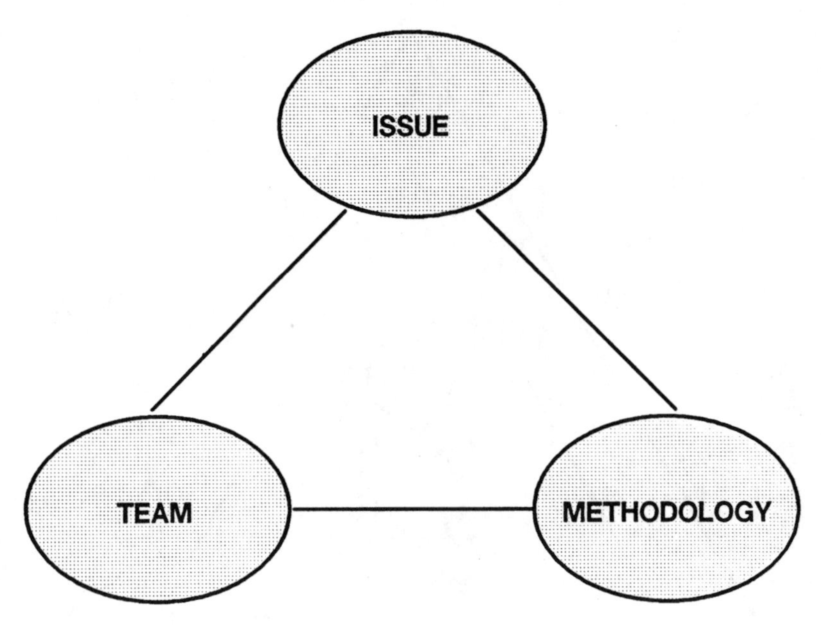

Figure 8.1 Basic elements of an issue.

study is or is not. Who funds the study and how much funding is required? Who approves the study? Who is the requestor? Who pays for implementation of recommendations? What are the study start and completion dates? Who should be on the study team? Who are the intended users of the study results? What is the study deliverable? What is the expected format of the deliverable? What levels of the organization should be involved? What geographic areas? What are the expected time, manpower, and cost? How do we introduce QFD into the department and company? What politics are involved? Do power plays exist? Is turf threatened?

Although the QFD process is a rational, systematic, and structured process, its foundation is structured around the effective use of people in teams and subsequently their interaction with the management chain in making recommendations that ultimately will lead to implementation and change. Once we add people, teams, organization structure, emotions, and power to the process, the process becomes more complex. Consider the following circumstances and roadblocks usually encountered in the QFD process.

1. QFD is always performed with interdisciplinary teams. Teams can waste time, be overly conservative, avoid decisions, and prematurely solve unclear problems.
2. Individuals involved in QFD usually have other full-time jobs and are already busy.
3. Strong parochial interests are common.

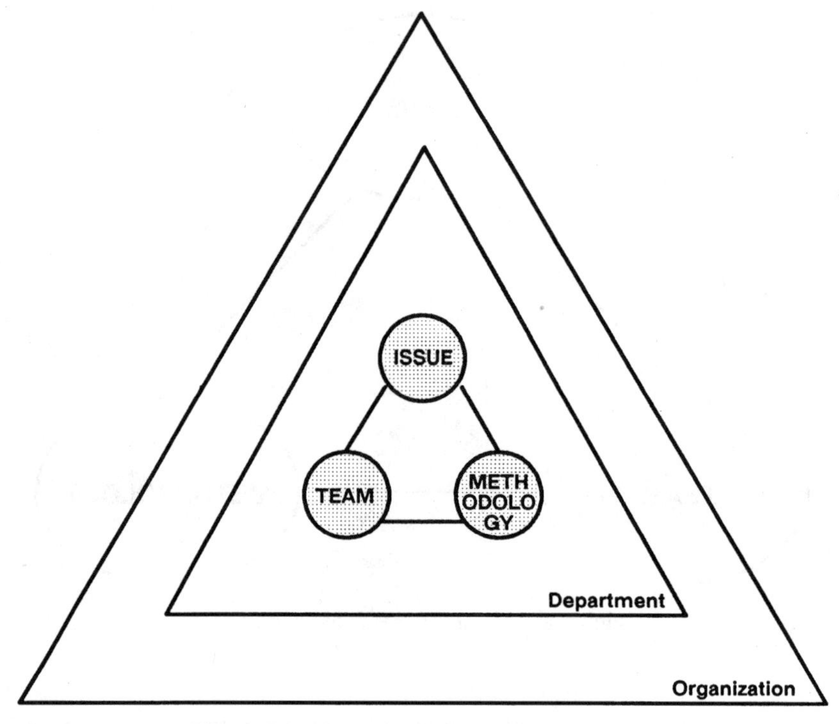

Figure 8.2 Organizational model of an issue.

4. The output of a QFD study may be threatening, especially to designers, planners, and decision makers.
5. Emotional as well as rational conflicts of interest are often generated.
6. The purpose of the QFD project is not always clear.
7. The final decision maker is not always obvious.

The success of the QFD project is enhanced if organizational, political, and behavioral aspects of the project are addressed early in the project. Roadblocks can be anticipated and planned for before they occur. The process facilitator should emphasize to the team that they should consider the first team meeting as the first day of implementation. The following "getting started" topics are particularly germane to the process.

INITIALIZING THE PROCESS

When starting a QFD project, the following subjects must be thoroughly addressed: purpose, alignment, scope, time horizon, completion date, market, study team members, assumptions, implementation, and project selection. Discussing and documenting these topics is one of the most important parts of the QFD process. A poor job here can cause teams to be off course, lose time, and develop excellent recommendations on the wrong thing. The QFD team should take as much time as needed to do a good job documenting these topics. Each topic will now be discussed separately.

I. Purpose:
- A. Why are we doing this project?
  1. To produce a revolutionary design?
  2. To produce an evolutionary design?
  3. To redesign an existing product to:
     - (a) Reduce costs?
     - (b) Increase productivity?
  4. To increase customer perceived value?
  5. To improve manufacturability? Assembly? Setup?
  6. To reduce cost only?
  7. To improve quality? Reliability?

Answers to these questions are not always obvious. Considerable amounts of time have been consumed by teams in generating a purpose statement. Sometimes it is necessary to check again with the requestor of the study to determine the purpose. It is better to do this up front, before the team works on the wrong objective. The verbal interaction necessary to draft the purpose statement helps bring focus and quickly gets team members working together.

Two basic questions must be considered when writing a purpose statement: (1) What are we trying to do for our customer? (2) What are we trying to do for ourselves (the company)? The answers to either of these questions cannot be at the sacrifice or expense of the other. Discussion of these two questions brings focus on the purpose of what specifically the team (project) is trying to achieve. The QFD team has a task to accomplish. This task is usually to improve some product/service in a way that meets customer and manufacturing needs and that provides some competitive advantage over some period of time. A purpose statement should be a short, broad definition of what is to be accomplished by the team and the QFD project and why. It essentially creates a target or goal for the team. Sample purpose statements might be: To manufacture component *x* for $1.00 for delivery by 199x. To reduce the total elapsed time required to implement a design change. To develop product *x* with a revolutionary design in a way that cost is less than $1K dollars and product is easy to use so that product has high appeal for market segment *y*.

The time spent in defining purpose is very important. For example, we recall a QFD project where the stated purpose was solely cost reduction. So much emphasis was placed on cost reduction that the team forgot or at least did a poor job of defining user/customer musts, wants, specifications, and constraints. The team did a tremendous job in reducing costs. Several weeks later during the implementation phase, the team discovered they could not implement most of their recommendations because they would jeopardize quality and reliability. It's true, they did not do a good job in the information phase. The purpose (cost reduction) consumed so much attention and mental energy that it jeopardized the rest of the study.

II. Alignment:

A. Who is the decision maker?
   1. Who is the person who will approve or disapprove team recommendations?
   2. Who pays the implementation cost?
   3. To whom will the team present an offer they can't refuse?

It is surprising how often teams have a difficult time answering these kinds of questions. Who the decision maker is, is not always clear. The decision maker must be identified; otherwise, the team can do an outstanding job on the QFD project only to make recommendations that some decision maker never wanted. A decision maker in the context of the QFD process is the person(s) to whom recommendations will be given, and who can approve or disapprove them. The decision maker may or may not pay for implementation of the recommendation. Sometimes there is a chain of decision makers. In this case, the first person in the decision chain that can say no is considered the decision maker and the person whom the team must keep informed.

The decision maker should be periodically informed of team progress and direction. In this respect, the decision maker should be used as a resource to the team. Through the interaction of the update sessions, the decision maker builds ownership in the team project. This, in turn, builds alignment so that team and decision maker are congruent in their content, direction, and expectation. Alignment increases the chance of acceptance of team recommendations and their subsequent implementation, and reduces the chances of the team performing an academic exercise.

Experience shows that decision makers don't like surprises. Surprises generally are threatening to a decision maker and increase the chance of a veto. This can happen when team recommendations are radically different from status quo or when they make historical decisions made by the decision maker look bad. Therefore, identify the decision maker and start building ownership and alignment as soon as possible. The quickest way to get started in stake building is to review the team-generated purpose statement with the decision maker. The more decision makers are consulted or interact with the team, the more ownership they will feel in the team project. By the time the team is ready to make its recommendations, the decision maker already knows what to expect. There are no surprises. There is a feeling of ownership. It is an offer that is difficult to refuse. The likelihood of approval is very high.

III. Scope:

A. What is included in the study?
B. What is not included in the study?
C. What is/is not the team able to control/change?
D. What are the boundaries within which the team operates?

It is important to bound the study so that it will be a manageable unit. The amount of time the team spends on scoping should not be jeopardized by frustration to prematurely start the QFD process. The danger of improper scoping is that the team may perform an excellent job and develop an excellent design for the wrong thing! Proper scoping increases the chances of the team doing the right things right. One model used in scoping is the black box input-output transformation model (see Figure 8.3). The black box is where some transformation or action takes place. It is the area of team responsibility. The inputs to the box are outside the scope of the team to influence. The inputs are given and are characterized by certain parameters. Coming out of the black box is some output produced by the transformation. Once this output leaves the boundary of the black box, it is no longer in the team's sphere of control. So, the scope of the team's project is everything within the boundaries of the black box.

For more complex studies it may be necessary to conduct a preliminary evaluation of the scope itself and to modify the original, if necessary, to develop a better understanding of what is really wanted. Sometimes money or time constraints influence the scope and restrict the boundaries of the QFD project. Reviewing the scope with the decision maker early in the project is time well spent.

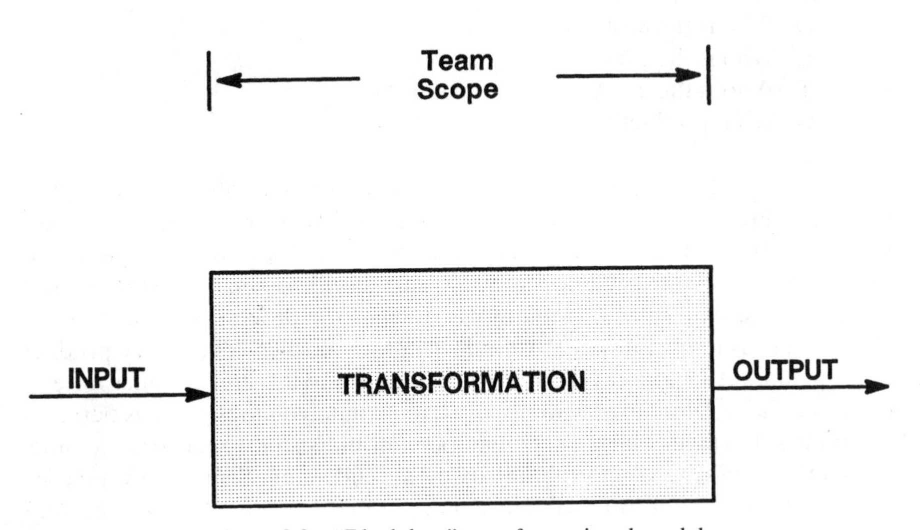

Figure 8.3 "Black box" transformational model.

IV. *Time horizon (consumer-type products):*
   A. What is the introduction or implementation year (date) for the product under study?
   B. Will there be more than one model (design)?
      1. Will the models be time-phased for introduction?
   C. Which model (design) does this current QFD study consider?

Too often teams get frustrated because there are so many models to work on. Many models or design variations are dependent upon when they will be used or introduced. Input from upper management is necessary to establish model type and introduction dates. That is, management should select the product and introduction dates; the team selects the functions and features to be studied. It is obviously necessary to have proper product/model/date selected before beginning the QFD process.

V. *Completion date for QFD process:*
   A. When must the QFD project be completed?
      1. When will the team present recommendations to the decision maker and other management?

Establishing an end date for the QFD project brings focus for completion. This, in turn, helps establish how often and for how long the team should meet.

VI. *Market (for consumer-type products):*
   A. What country? Domestic only?
   B. What market?
   C. What segment?
   D. Who is the customer?
   E. Who is the user?
   F. Who is the chief buying influence?
   G. What product?

Answers to these questions will shape the outcome of the entire QFD study. Product features and their importance will vary greatly across market segments. Segment also determines what products (both our company and other manufacturers) are considered. This can be a time-consuming and sometimes difficult task. A popular way to start the process of defining market segment is to establish a Pareto distribution of sales volumes by product line and market segment. When doing so, it is best to use unit volumes instead of sales dollars. There are various ways to correlate volumes, such as percentage units sold versus market segment, percent units sold versus product line, product line volume versus market segment, and so on. Signals to watch for are things like 80% of total volume coming from one segment or 80% of total volume coming from one or two product lines. The team and project time should be spent on the vital few versus the trivial many. The team must determine whether volume is the correct indicator to establish which market segment to work on. Should the team be working on the high-volume areas to design or redesign a product to maintain market share? Should the team be working in low-volume areas to design a product to capture more market? There are many approaches to determine which product to work on. The final approach will depend upon product, market, management, and the line of business strategic plan. Sometimes market or segment is dictated by upper management.

For consumer-type products it is, many times, not as simple as it may appear to define market, segment, customer, user, and *chief buying influence*.[2] It is important to define who these people are because they can each have their own set of needs. Tom Cook, president of Thomas Cook Associates, offers a good example (Reference 2, p. 147):

> Consider a baby food product. The baby would be the user. The mother or father would be the purchaser or customer. However, the chief buying influence may be none of these, but would instead be the pediatrician. . . .

So, for whom are we designing the product? The resolution of customer, user, and chief buying influence must be established before developing design parameters. Market and customer are discussed in more detail in Chapter 9.

VII. *Study team members:*
   A. Based on the answers to the preceding six questions, are the right people on the core team?
   B. Will ad hoc members be needed on the team at certain times in addition to the core members?

When a core team is assembled, the members will remain on this team until the project is completed. Generally the core team consists of a minimum of the chief designer, the project leader, a QFD process facilitator, and the QFD team leader. On numerous occasions, study teams discovered that certain key people are not represented on the core team. The missing representatives are discovered only after the teams have discussed the previous six questions. Usually the proper people are already on the core team, but additional ad hoc members may have to participate for a limited duration at appropriate times in the future. These people will have specialized knowledge for certain areas of the project. Making them permanent members of the core team would not be practical (Figure 8.4).

Team members must represent both the customer (user) and the producer. This would suggest the following types of disciplines: design, manufacturing, operations, marketing, customer service, user/equipment operators, manufacturing engineering, product planning, and so on. When choosing team members, you should focus on a person's expertise and not on his or her position in the organization.

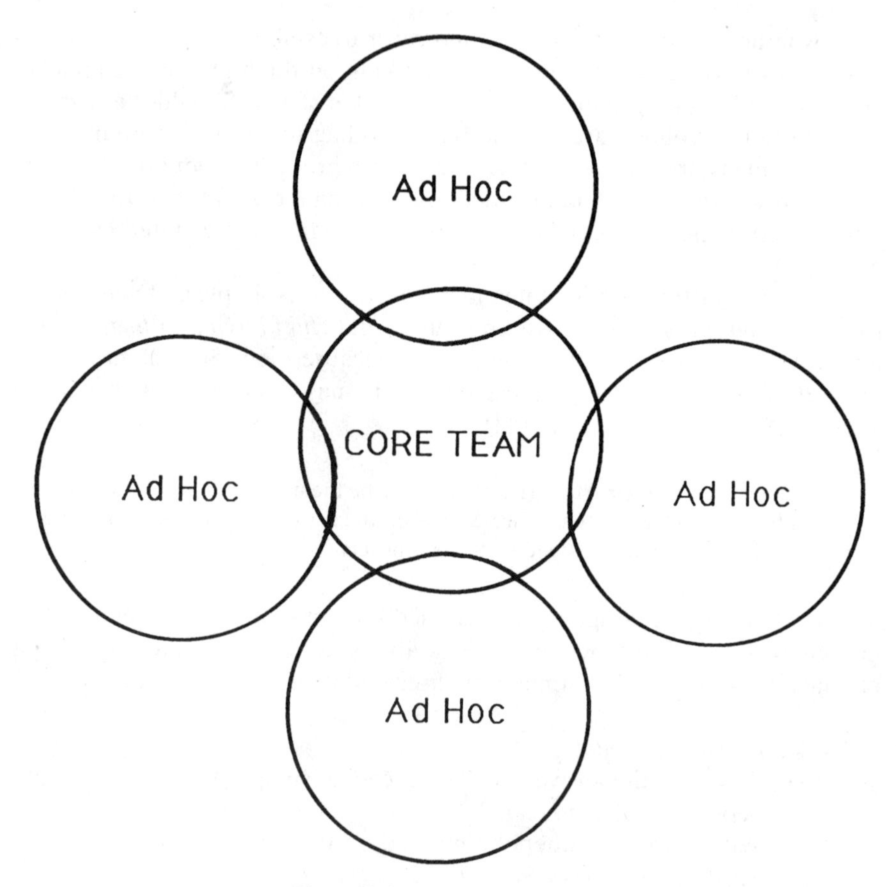

Figure 8.4 Core team model.

A QFD team normally comprises three to seven members. With more than seven members, interaction becomes complex, discussions become indistinct, and the group begins to fracture. An odd number is also helpful, because it reduces chances for split decisions. The following characteristics are very important when assembling a team:

1. It should be interdisciplinary, incorporating a good balance of background, viewpoints, and disciplines, as well as good geographic representation.
2. Members should be from equivalent levels in the organization's hierarchy in order to minimize peer pressure and politics.[3]
3. It is sometimes helpful to include a decision maker on the team, because acceptance of results may depend upon who is on the team. Caution should be taken, however, if having the decision maker on the team. His or her presence may induce peer pressure, and if the decision maker assumes a boss role, this may inhibit participation and candid discussion.

4. It is necessary that one or more members be versed in the QFD process. Alternatively, a third-party facilitator or outside consultant can supply the QFD methodology.
5. At least one member must be an expert on the product or subject being studied.

The team members themselves should

1. Be at least generally familiar with the product or area of study.
2. Know the sources of data for their area of expertise.
3. Have interest, motivation, and commitment to engage in the task.
4. Be able to get cooperation and assistance while representing their organizational area.
5. Have sufficient time to do the job and be engaged long enough to provide continuity to the product.
6. "Be able to create, accept, and be eager to exploit change."[4]
7. Have an open mind and be able to work and communicate with others in a team environment.

Team formation and team member selection should be directed toward creating an interdisciplinary team as opposed to a multidisciplinary one. Figure 8.5 illustrates the difference between the two definitions. Generally, the terms are used interchangeably. Making a distinction between the two terms helps to emphasize the importance of working together to do a high-quality job and to provide commitment and energy to make something happen. A good example of a team is that of Eric Berne's.[5]

Referring to Figure 8.6, the outer circle represents the membership boundary. This boundary encompasses the team itself, a selected group of people. The outer circle could be considered as representing a QFD team. Internal to this team is another boundary or core that represents the leader. This core also consists of procedures and rules (Figure 8.7). The leader directs the team through a certain methodology or constitution. The process, or methodology, forms the framework for the leader. What should characterize a QFD team is that it should not be a personal leadership role. There must be a coordinator, someone who sets up the meetings and the agenda. For the team to really be effective, the personal leadership must diminish as time goes by so that the leadership is shared by all of the team members. This is not possible unless there is a substitute for the leader, some sort of core leadership that comes from constitutions, methodologies, or rules. The QFD process is such a methodology. In the case of a QFD team, it is the QFD techniques and methodology that fulfill the leadership role. The leader dissolves into the team so that the team members carry on through the structure of the QFD process.

The various meetings are patterned on a plan. The team devotes the first several sessions to gathering information about the problem and defining it in

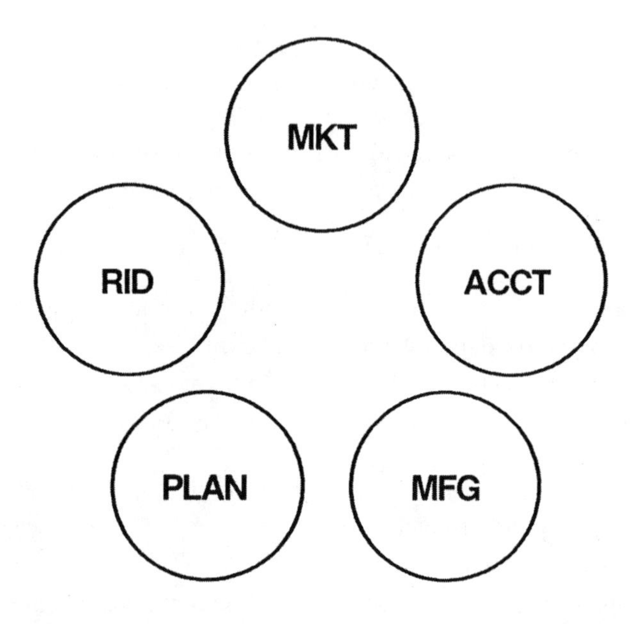

MULTIDISCIPLINARY

INTERDISCIPLINARY

Figure 8.5 Team cohesiveness comparison.

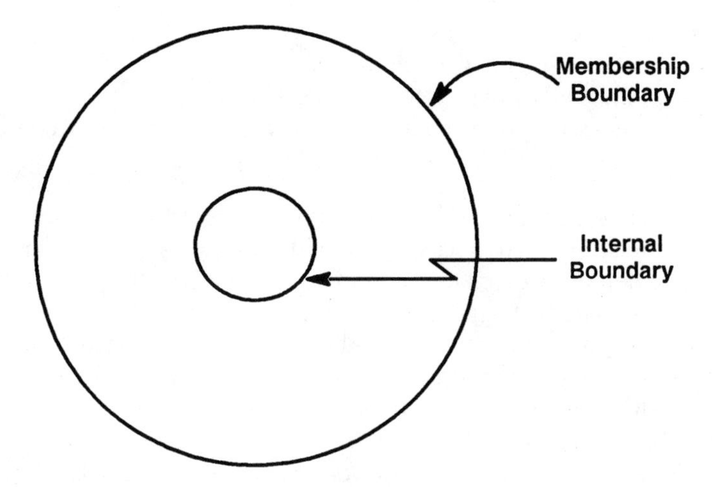

Figure 8.6 Team boundaries.

a better sense. The team then devotes more sessions to getting ideas how to change or improve the particular situation that is presenting the problem. It may seem trite or ritualistic, but in a system that has to function without a charismatic leader—and it must do this in order to get the team members committed—there must be some sort of goals or procedures inherent in the system. It is these goals and procedures that form the basis for carrying on. They elicit commitment from the members and provide the vehicle and momentum for overcoming obstacles. They are an operational method of putting the ideas of behavioral science to work. Other excellent references on teams are Fiorelli,[6] Dillard,[3] and Parker.[7]

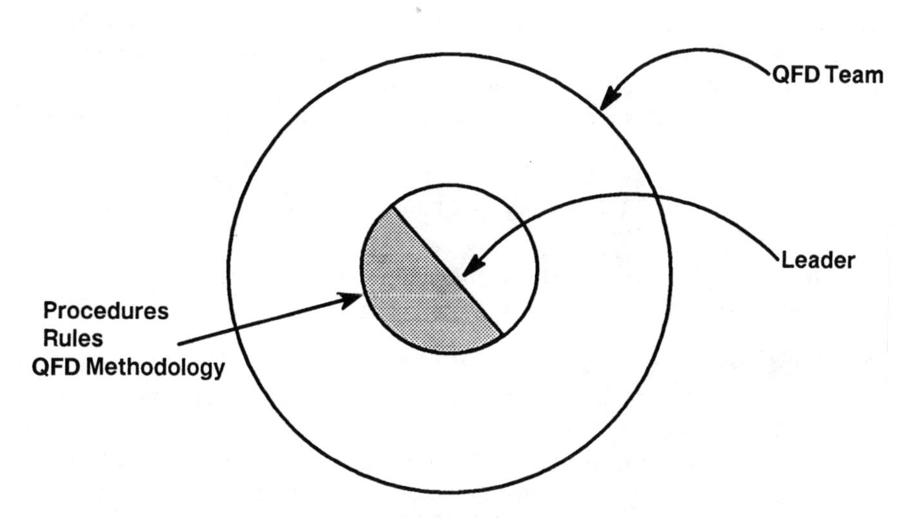

Figure 8.7 The methodology-team balance.

VIII. Assumptions:

A. What are the initial assumptions?
 1. The study will proceed based on these.
 2. As assumptions change or become fact, the study content and direction will be altered to reflect the changes.

All projects begin with a basic set of assumptions. Too often these assumptions are not documented or visible. Many times they are taken for granted. Forcing documentation of assumptions in the beginning of the project enhances communications and highlights gaps in information. It can also assist as a double check on the purpose statement. Assumptions are things like "the building will be occupied by the shops division"; "the building will have two floors"; "we will have in line-testing"; "we will use the current power source"; "branch offices will be located in areas *x* and *y*."

To keep assumptions visible and up-to-date, it is good practice to write them on a chart pad. This pad is then posted on the wall of every meeting. As new assumptions arise, they are added to the list. As current assumptions are confirmed, they are so noted. This on-going list is commonly referred to as a parking lot. Parking lots are also used for other subjects as well, such as premature ideas, questions, and action items.

IX. Company business plan.

It is surprising how often teams design products without knowing the business plan or the strategic intent of the company or line of business. Too often design teams never think to ask for this information. The worst cases are those where lower ranking individuals are not allowed to see the company business plan. The team is directed to design a product and upper management will decide whether or not the team's recommendations fit the need of the company. Such noncommunication will surely be short-lived and company resources will be needlessly wasted. This is the price companies pay when individuals guard information and consider it as power.

To hopefully avert such a situation, the following questions should be answered:

1. Does the team have a copy of or have access to the company business plan?
2. Do the answers to the first eight questions fit or reflect the company business plan?

X. Implementation:

A. What are the necessary steps to implement recommendations resulting from the QFD study?
 1. What factors could prevent implementation of recommendations?
 2. How can these opposing factors be eliminated or resolved?

B. Consider the first day of the QFD study as the first day of implementation.

Discussing these issues and their countermeasures early can increase the chances of success as well as help shape the study itself and the process. As an analogy, an artist must visualize what the final painting or sculpture will look like before she begins the work. So, too, the QFD team should visualize the possible proposals and recommendations that may result from the QFD study. Based on this vision of the results, the team can consider possible solutions to sources of opposition. These sources of opposition may vary with time. The method of force field analysis readily lends itself to this discussion. Basically, the team considers the approval process. It considers how opinions of management can be determined without risking a pocket veto. Other things considered in gaining commitment are: Who implements? Who budgets for implementation? Who will be affected? Who gains, who loses? Who can delay or block implementation? The answers to many of these questions begin to surface during the team discussions on alignment mentioned in item II. Discussing the commitment issue early in the process is a step toward building implementation.

XI. Project selection.

Many QFD studies arise out of necessity in a very specific, well-defined area. Consequently, a predefined project may obviate the need for formal project selection. However, the resources that can be allocated to a QFD study are limited. General criteria that should be considered are:

1. Be a solvable problem. The need should be real and should be supported by management.
2. Have a good probability of success and implementation.
3. Have objectives that are credible.
4. Be important to the people in the area being studied.
5. Have the commitment of the requestor and the team members.
6. Have receptivity. The sponsor or decision maker must be receptive to change.

A U.S. Department of Defense publication[8] lists additional project selection criteria. There will be other technical parameters necessary for project selection. Many lists of project selection criteria abound in the literature.[9,10] The above list is generic to all projects. This list of items should be kept in mind when selecting projects, because too often supervision or management wants to apply the QFD process to the most pressing problem, which many times is also the most difficult.

If QFD is being introduced for the first time, it is wise to start with an easy problem with virtually a guaranteed successful application of QFD, a problem where you are sure recommended changes will be implemented. Build credibility first, then take on the tougher projects, because the first project establishes the pattern and reputation for all succeeding projects. Sacred cow products/projects should be considered cautiously, if at all, for project selection. What is a sacred cow? It is a product that has been around for a long time and was most likely invented years ago by the current president or other higher officer in the company. It is that person's pride and joy, even though it is overdue for updating and cost reduction. Usually, any time spent investigating such a product results in roadblocking and discomfort if the inventor/officer has less than an open mind. Other products that may have a propensity for project failure are those that have high esteem value or those that have been previously extensively investigated. Fowler[11] recommends that the product or project selected should be the heart of the line with maximum potential spin off to other models in the product line.

LAUNCHING AND CONDUCTING THE QFD PROJECT

Third-Party Facilitator

The behavioral and scoping exercise to initiate the process can be complex. The entire process from launch to finish is best done with a neutral third-party process facilitator to lead team building and process. The use of a third-party facilitator also helps to keep honest people honest. Otherwise the team may get mired in turf protection, drown itself in detail, get lost in the woods, or rush to solve problems quickly and shortcut the process. The process facilitator provides the QFD working model and keeps the team on track.

From my experience, most people involved with QFD projects are used to getting problems solved quickly. Usually in the course of their jobs, they are given little time in which to resolve any given problem, and so they are used to reaching a conclusion quickly with little data. Consequently, the QFD process, which takes more time and requires more data, will be frustrating to these types as team members. They need to be reminded and reassured that this frustration is normal and that answers and solutions will evolve. However, with the QFD process they will have more data and ideas, which will enable them to make a better, more informed decision. In fact, the hours spent on the QFD project will probably be the most time the team members will have worked together in their career on a common problem and objective. They and the decision maker should feel assured that the team will have considered all possible solutions before making the final recommendations. The third-party facilitator can play an invaluable role in the dynamics of team interaction in its journey to delivering final recommendations.

Meetings

Meeting length and frequency are also important. The initial launch meeting should be scheduled for a full day. Regularly scheduled follow-up meetings should be scheduled at the launch meeting. These should be four hour meet-ings held at least once per week until the QFD project is finished. Because of the complexity and detail in designing products, infrequent meetings less than four hours long do not work well. Team members spend too much time getting back to speed. So, at the onset of the project, the team is told to expect one or two meetings per week. Each meeting will be four hours in length to provide continuity and will occur at regular standing times and days (e.g., every Tuesday from 1:00 to 5:00). Team members can expect to spend 40 to 60 hours of team time over the next 2 to 3 months. By now, not only the team members but you, as well, are probably wondering how all of this is going to happen. After all, this many people for this amount of time! It is very beneficial to have management endorse the QFD study in a properly distributed letter to the supervisors of each of the team members. The letter should define the project, the sponsor, the scope of the project, and the priority. Most importantly, the letter should designate the team members, give them the time and priority to participate in the study, and pave the way to access needed information. Refer to Figure 8.8 for an example of such a letter.

The Launch. As mentioned earlier, experience shows that the initial launch meeting should be a full day. It should be kicked off by the decision maker or management person requesting the study to lend more credibility to the project and state how important it is to the company. For some team members, this may be the first time they will have had to interact with this level of management. A good gesture, and money well spent, is to have a company-paid lunch and/or an evening cocktail party to celebrate the initiation of a very important project. The location of the launch meeting is also important, and experience shows it is best to locate it off-site at a neutral site away from telephones, desks, and tendencies for team members to skip out to address external issues. If team members are easily distracted, the process facilitator and the team begin to lose control, which in turn prolongs completion time. Team members who are present become discouraged. Local motels, hotels, or conference centers are excellent choices for this occasion.

Launch Agenda. Launch agenda, pace, and atmosphere are very critical to a good start to set the feeling and engender commitment to the project. A time-tested agenda is illustrated in Figure 8.9. The decision maker introduces the project and stresses its importance and emphasizes how critical the team is to the project's success.

Another point to mention here is that the organization competes externally and not internally. The decision maker then introduces the QFD facilitator and team leader. A team warm-up exercise is used to draw people in if team members do not know each other. One that we use is the "who am I?" exercise, which requires each team member to prepare answers to the following questions: Who are you? How do you like to be addressed? How do you not like to be addressed? Where do you work? What do you do? What do you like most about what you do? What do you like least about what you do? How do you feel about today? Tell us one interesting thing about yourself that

DATE

TO: W. Johnson
G. Smith
W. Rowe
R. Williams

FROM: I. M. Boss, Anywhere, Rochester, N.Y.

SUBJECT: Wigit Assembly QFD

The Wigit Engineering Section is sponsoring a QFD project to determine a least-cost alternative for a new Wigit subsystem design.

In this endeavor, we have assembled a team comprised of the following members:

1. Anthony Brown, Processing Engineering
2. Michael Rhodes, Product Development
3. Edward Jones, Industrial Engineering
4. Carl White, Customer Equipment Services
5. Mary Melone, Manufacturing Engineering
6. Charles Nelson, Reliability Engineering

Because of their knowledge and experience, I am asking if you would approve the above individuals and necessary time to participate in the project. It is expected that this may require four half-day sessions during the next four weeks. This time includes attendance at meetings as well as possible homework between meetings.

The QFD process has a proven track record on other projects both internal and external to the company, and we believe the effort spent in applying it to the Wigit assembly will result in significant cost reduction and design improvement.

I appreciate the fact that your people are very busy with other assignments, and it is with this in mind that I am asking for your cooperation. This endeavor will provide timely input to new product development activity. Thank you for your cooperation.

Figure 8.8 Management letter announcing the QFD project.

might be useful to the other team members in getting to know and work with you.

The members are given five minutes to prepare the answers. Then each member presents his or her answers to the group. Only five minutes are allowed for each individual presentation. Next, the 40 minute video by Joel A. Barker[12] titled *Discovering the Future: The Business of Paradigms* is shown. This is a very relaxing video about looking into the future and the effect that

AGENDA		
DATE		
	WELCOME	
8:10	Kickoff	Decision maker
8:20	Who am I?	Group
8:45	Video	Group
9:20	Discuss video	Group
9:35	Teams	Facilitator
9:40	Responsibilities and operating guidelines	Facilitator/group
9:50	Overview of process	Facilitator
10:15	Break	
10:30	Getting started	Facilitator/group
11:30	Lunch	
12:30	Begin process—Customer needs	Facilitator

Figure 8.9 QFD launch agenda.

paradigms have on blocking our vision. A short paradigm workshop (approximately 15 minutes) is conducted. The following types of questions are discussed by the team with process facilitator moderation:

1. What specific paradigms have an effect on our potential team output?
   (a) Positive effect?
   (b) Negative effect?
2. What are some paradigms that influence the way we think?
3. What are some paradigms that can influence our designing a product and getting it launched successfully?
4. What are some paradigms that can affect progress on this team project?

The facilitator writes the group answers and ideas on chart paper. The team may highlight the one or two most significant paradigms. One option at this point is to have the team discuss how it might minimize the effects of negative paradigms and/or maximize the effects of positive paradigms. Experience is showing that negative paradigms are the first to surface. The process facilitator needs skill at this point to keep the discussion from becoming a doomsday exercise in futility. Remind the group that negative paradigms, like problems, can be interpreted as opportunities. Generally, the list of paradigms is posted on the wall at this meeting and all subsequent meetings to keep the team members alert to paradigms that are affecting progress.

During the paradigm workshop, team members quickly become interested and involved. They become aware of and sensitized to paradigms. They recognize situations where paradigms are influencing their outlook or restricting their thinking and creativity. They also are alert to other team members' paradigms. In an extreme case, surfaced paradigms may redirect an entire project, if the project purpose and/or assumptions are riddled with false paradigms. Surfacing paradigms can sometimes have stunning affects.

After the paradigm workshop and discussion, we give a short lecture about team dynamics and the stages of team development from immature and ineffective to mature and effective teams. Many times it is necessary to have a short team-building workshop as part of the QFD project launch. We use a retrospective planning exercise. First we get the team to describe the ideal future state they are trying to achieve. Next we assume that this future state or goal has been achieved, and ask, how did we do it? What characterized the team's effort and the members' ability to pull it off? What contributed most to success? Answers to these questions are posted on chart paper. The team uses Pareto voting to narrow down the list to the most important elements. Next the team determines if any gaps exist between now and the desired future state. They then determine what opportunities they have to fill the gaps. The output of this team-building exercise is a set of team principles that are going to be used to fill the gaps and help get to the future state. Quality of team output and commitment are then discussed.

Throughout the paradigm workshop and the team-building exercise, it is necessary to stress and reaffirm that the competition exists outside of the company and not inside of the company or the team. Many times it is useful and fun to have the team select a team name. We have even given prizes or mementos such as team T-shirts or sweatshirts embossed with the team name.

The following responsibilities and operating guidelines are reviewed.

Process (QFD) Consultant Responsibilities and Guidelines

1. Provides process model only.
2. Keeps the team on track.
3. Is not a content expert.
4. Will not "carry the team's monkey."
5. Will not provide all the energy to do the project!

Team Responsibilities

1. Will produce a deliverable (recommendations).
2. Will communicate with each other.
3. Agree to disagree.
4. Make decisions as a group.
5. Communicate as a group.
6. Respect each other's views.
7. Actively listen to each other.
8. Keep an open mind.
9. Focus on "what is right," not "who is right."
10. Won't be afraid of conflict and will use it to the team's project's advantage.
11. Will have fun!

It is really best, if time permits, to have the team develop their own list of operating guidelines or vote on the most important from a prepared list such as above. Discussing these issues brings focus among all individuals involved. An overview of the QFD process is then given to prepare the team for what they will be doing and the process they will be using. After the coffee break the following "getting started" questions are answered in detail:

1. Why are we doing this project?
2. When is it to be completed?
3. Who is the decision maker?
4. What is the scope? Include? Not include?
5. What country, market, segment, user?
6. What type of product? Evolutionary? Revolutionary?
7. Do we still have the right members on the team?

These topics have been discussed in more detail earlier in this chapter. Answering the questions starts to bring the team together to focus on the task. Answers to these questions may take longer than the process facilitator originally allows. However, the questions must be answered! So, flexibility is one important trait a process facilitator must have.

The team is now ready for lunch. After lunch or after the previous questions have been answered, the team begins the QFD process by focusing on customer needs. This will last the rest of the day. Fifteen minutes are allowed at the end for team/project "administrivia." This is where future team meetings are blocked on the participants' calendars (which they were told to bring to the meeting).

All future team meetings will continue to follow the QFD process. Members quickly become involved in the QFD process to the point that the process becomes the vehicle for dialogue to carry the project to completion. Teams can work effectively or can waste time and make poor decisions (or none at all). The structure and momentum of the team is maintained by the QFD process, methodology, and techniques. It provides the vehicle for overcoming obstacles. It helps to encourage rational conflict and minimize emotional conflict. It helps the team work at the right level of detail. To check the health of team meetings, commitment checks, or references to how well the team is following the team guidelines are very helpful to the facilitator and team to maintain human dynamics.

Team-building workshops conducted prior to doing the actual QFD project are very effective and well worth the time. The purposes of the workshops are: (1) to help the members develop communications skills and learn to work together as a group, and (2) to use the group skills effectively to analyze problems, make decisions, and work as a team. Unfortunately the human relations/personnel relations practices in most companies do not support teamwork and social skills, nor are people with those skills necessarily the ones hired. There are some excellent sources on team building and process consultation. See the categorized list of references at the end of this chapter.

Another format for a team-building workshop is to take one half a day immediately before the QFD project to have a lecture/overview on QFD. Following the overview, the rest of the day is spent building a house of quality (HOQ) on a simple subject like a pencil or a three-hole punch. In order to finish in the remaining time, the matrix size is limited to 6×6. The next morning the team knows how to build a HOQ and is ready to roll.

DEVELOPING A QFD PROGRAM AND ACTIVITIES

The accelerating pace of change that is engulfing all organizations today is making planning, both reactive and proactive, a crucial process for survival. QFD can be used to develop and improve the input needed to make a more informed decision. To do this requires that QFD be established for long-term success. In this regard: How does one get a QFD project started? How is it introduced into the organization? Where does it belong in the organization? How does one put the QFD plumbing in place? In considering these questions, the essential ingredients for integrating QFD into an organization are top-down management support and participation, and line and staff support across organization functions.

Top-down management support is not enough. Management, at the appropriate level, should actually participate in the effort and activity. Let's face it, management attendance at one or several QFD team meetings can go a long way. The team can ask questions first hand, hear the latest information, as well as ask/receive course corrections.

Horizontal support is also necessary because QFD interdisciplinary teams can cross many functional areas, all of which must buy in to make the project, as well as a QFD program, successful.

In addition to top management support, there must be a QFD champion to drive and nurture the effort. Simply put, the company must designate a chief plumber who is the focus and mastermind to lay the plumbing in place. In this respect, it is also important that the project and/or team leader be a champion of the QFD process. This project/team leader should have a reputation for implementing decisions and getting things done. It has been my experience that the strength of QFD and QFD applications is proportional to the respect for and effectiveness of the project leader.

Where does one place QFD in the organization? The first function that comes to mind is to place it in the quality organization or department. This, in my experience, has worked very well. I get the same feedback in talking to many colleagues from other companies. Two other excellent choices are to locate QFD in the value engineering department or the industrial engineering department. People in these departments work across many functional areas and are well known. They also know the climate in which QFD is to be used and can help in implementation.

How do you get QFD started? Generally a management overview, less than one hour, is given. If there is further interest, similar overviews would be given at lower levels. These overviews should be started by an upper management introduction. Many times management will also select a product to be designed or improved using QFD. This becomes the first pilot. Chapter 7 contains more details in starting a QFD project. In any case, you should start with one project and this project should be small and easy to do with a guaranteed success and implementation. The other projects will ride on the success of this winner.

How do you start training? One approach that is *not* recommended is mass blanket-coverage training of everyone. This has connotations of the "program-of-the-month." Too often people trained en masse don't use the course contents until a long time after training. Chances are that by this time they will have forgotten what they learned and will have to be retrained. The most effective training has been just-in-time (JIT), where it is learned on-the-job as part of a QFD team/project.

I cannot reiterate enough that there is one prerequisite that all QFD practitioners, managers, and consultants would unanimously vote as still being the most important: The QFD program must have complete top-down management support and involvement. I realize that each company has its own personality and culture and that QFD programs have to be tailored to each. The preceding comments and suggestions have led to many successful applications and can be considered as a seedbed for further ideas.

OTHER FACTORS AFFECTING APPLICATION OF QFD

Group/Team Dynamics

The most successful QFD teams I have worked with were those that:

1. Had clear goals and knew where they were headed.
2. Had the freedom to work their own way without considerable interference (there was no punishment for failure).
3. Had free access to all needed data.
4. Had full responsibility for their deliverables.
5. Had a good balance between rational and intuitive thinking

Finally, those teams that were able to respond most quickly and implement changes and recommendations the fastest were those teams whose members were empowered to make decisions without bureaucratic interference. This is extremely important because those companies who will be successful in the future are those who can respond the fastest to customer needs. Time, not quality, will be the battlefield. Team response and effectiveness provide the foundation for the company's success.

Benefits

Let us start first on the positive side with the benefits of using QFD! If we can document these, they, in turn, can be used as factors to develop a positive mind-set for applying QFD. First of all, why do QFD at all? So that:

1. We don't design something that is not needed.
2. We design what the customer wants.
3. We communicate better, internally/externally and vertically/horizontally.
4. We can minimize redesign and engineering changes.
5. We can build it right the first time.
6. We can improve design and reduce costs.
7. We can challenge and set realistic specifications.
8. We can manage costs on purpose and remain competitive.

This list is a beginning. You can certainly add more to it. The point is that these items are factors affecting application of QFD. Too often, though, they are overwhelmed by negative paradigms that are strongly entrenched in the company.

What are the benefits to the company?

1. Increased communications, vertically and horizontally.
2. Higher quality at lower cost.
3. Better value to customers; reduced value mismatch.
4. Simpler manufacturing methods.
5. Faster assembly.
6. Better material selection.
7. Just-in-time type inventories.
8. Shorter development and ship-to-stock time.
9. More realistic specifications.
10. Reduced postintroduction product problem solving.

This list could be expanded to fill many pages. The type of benefits accrued depend on where and how the QFD process is applied. In this respect, this list can also give you an idea where to apply the QFD process. However, a common thread benefit running through all types of applications is increased company communications and a more customer-oriented value-added product at lower cost.

What are the benefits to management?

1. A means to have their people use a structured, focused approach to reduce costs and value mismatch.
2. A means to set an example for their people:

   (a) In being proactive in cost management.
   (b) In facilitating communications.
   (c) In encouraging a more customer-oriented approach to achieving value-added designs.
3. A connecting channel to be more interactive with subordinates.
4. A structured means to share strategic intent and company direction.
5. A way to tap the potential of subordinates.

What are the benefits to the QFD practitioner or team member?

1. Personal development opportunity.
2. Ability to be seen by others as a creative individual.
3. Potential for high-level visibility and major impact.
4. Recognition from others.
5. Legitimate sounding board for ideas.
6. An opportunity to improve communications with colleagues, superiors, and clients.
7. An opportunity for another viewpoint toward design.
8. An opportunity to work interactively and in parallel with other departments rather than in a series.
9. An opportunity to view problems and opportunities from a total corporate perspective rather than shop floor tunnel vision.

Barriers

Even with all of the benefits of the application of the QFD process and a QFD program, it still can be difficult to get QFD started within a company. Some of the most frequent barriers to QFD are the following.

1. *Lack of Management Support.* Management often states they support QFD, but too often it turns out to be lip service. Many teams get formed with a designated amount of time allocated for the project. The allocated time usually competes with other activities having higher emphasis or priority from management. This results in teams having variable attendance, and lack of continuity on the project. A person will spend time on the most visible project, the one that commands the most attention, or the one emphasized by management or supervision.

2. *The Reward System.* Generally, in most companies, there are no rewards for using processes like QFD and COPC. Performance appraisals and pay raises are based on history. They are based on how well you satisfy things like schedules or sales quotas, or how many problems you solve. Problem solving and fire fighting provide the grist for making heroes. Problems

are usually well known internally and make problem solvers highly visible and well rewarded. This tends to encourage problem solving instead of problem prevention. People who prevent problems are often unknown and unrewarded. The QFD process is both problem solving and future oriented. When asked to participate on a QFD project, people are sometimes reluctant to do so, for fear of taking time away from working on their own problems and the real work they are rated against and paid to do. Also, because QFD is often future oriented, it can be perceived as threatening to historical elements of performance appraisals, which can discourage people from using it. Compounding this barrier is the fact that there are no consequences or reprimands for not using the tools. People are not held accountable!

3. *We Don't Have Time.* Team members are pressured to meet schedules on current projects and see QFD as a delay to these time goals. Benefits are not always visible at the beginning of a QFD project. It is hard for teams and individuals to risk their schedule attainment without supporting proof that they should apply the process. People also have a tendency to seek out quick solutions to problems, and when learning about a 40 hour QFD project, they develop considerable anxiety about participating. The time frame to final solution in a QFD project is not congruent with their personal shorter time frame for solving problems.

4. *Change.* Change never comes easily. Most people resist change, especially product designers, manufacturing managers, production engineers, assembly people, and plant management. The QFD process by its very nature creates change, sometimes drastic change. This kind of change represents a threat to the above types of people, because QFD seeks to change the product/process, which in turn, creates a cascade of changes down the line. The plant personnel are accustomed to managing day-to-day activities and immediate problems. This is the nature of their work and what they are rated on. By using QFD, managers and supervisors must take a longer term view, beyond the present. They aren't used to doing this and by doing so, they get overwhelmed with perceptions of change. They feel uncomfortable. The first tendency is to resist QFD.

5. *"We're Already Doing It."* Our immediate response to this barrier is: "Show us, we would like to learn from your experience!" Perhaps, in some respects, this may be true. Many individuals and teams have or are already using "QFD"-type data. However, these teams usually do not nave the data connected through some structure. They usually have random pockets of isolated data. Because QFD data, labels, and so on, look familiar, people believe they are using the process. This is a barrier easily converted into an attribute! All one has to do is help them get the QFD data into a QFD format. The relationships and interconnectedness are many times more valuable or useful than the actual data. It can be difficult to convince individuals to use the QFD process to structure and quantify their data.

Success Factors for QFD

1. There must be a perceived advantage to using the process. The key word here is perceived. Whether or not there is a real distinct advantage is not as important as the fact that the potential user perceives an advantage.
2. The QFD process should be perceived as compatible with the existing way of doing things. Processes that are different have many strange terms and acronyms and disrupt the way of doing business and consequently invade the comfort zone of individuals who mentally tune it out.
3. The process should be perceived as causing low anxiety and having a high comfort level. The process should appear to be "safe."
4. There should be little to lose if the process is terminated before closure.
5. The process must not be perceived as another "flavor-of-the-month."

CONCLUSION

Given the benefits of QFD and despite the barriers, a QFD program should be a must for a company to remain competitive in the future. QFD is currently quoted by many companies as being among their competitive weapons for the future. Those companies who successfully implement a QFD program will have a formidable advantage of customer-oriented, high quality, value-added, low cost product designs.

With the surge of total quality management activity, the designing and operating of teams, especially design teams, will be increasingly more important. The parallel team approach to product design and manufacture will become the standard way to function as opposed to the traditional throw-it-over-the-wall-to-the-next-guy serial approach.

Currently there are several terms that describe this parallel approach: *simultaneous engineering, concurrent engineering, multifunctional team design,* and others.[13] Using the parallel team approach will require many changes in organizational structure, roles, responsibilities, performance appraisals, and reward structures. Organizational goals and the team's purpose must be congruent. Empowering and staffing the team will become more critical to successful team operations. Changing the rewards to fit with concurrent engineering team design will present a real challenge. A few companies are now starting to tackle these concerns.

People and teams can complicate the QFD process but are a necessary part of it. Behavioral science aspects must be treated as carefully as any other part of the QFD process. Look ahead, predict the problems, and carefully plan to deal with them. Remember, the first team meeting should be considered as the first day of implementation.

HELPFUL READING

The following references are grouped by subject matter. The complete details of each reference can be found in the reference or bibliography section at the end of this chapter.

MANAGING AND ORGANIZING QFD

Author	Reference Number	Bibliography Number
Managing and Organizing QFD		
Barlow		2
Berne	5	
Marsh et al.		10
Riegle	4	
Sehr		17
Initializing the QFD Process		
Bachli		1
Boothe		3
Dillard	3	
Fiorelli	6	
Parker	7	
Riegle	4	
Rivet		13
Schein		14
Schein		15
QFD Project Selection		
Drozal	9	
Fowler	11	
Mudge		11
Parks	10	
U.S. Department of Defense	8	
Team Member Selection, Team Building, Process Consultation		
Bachli		1
Boothe		3
Fiorelli	6	
Fordyce		4
French		5
Gomolak		6
Kiefer and Stroh		8
Parker	7	
Port, Schiller and King		12
Schein		14
Scholtes		16
Welter	13	

REFERENCES

1. Warfield, J. N., "An Assault on Complexity," in *Batelle Monograph No.* 3, Battelle Memorial Institute, Columbus, Ohio, April 1973, Chapter 1, pp. 1.1–1.8.
2. Cook, T. F., "Determine Value Mis-Match by Measuring User/Customer Attitudes," *Proceedings, Society of American Value Engineers* **21,** 145–156, May 1986.
3. Dillard, C. W., "Value Engineering Organization and Team Selection," *Proceedings, Society of American Value Engineers* **10,** 11–12, May 1975.
4. Reigle, J., "Value Engineering: A Management Overview," *Value World* **3**(3), 4–8, 1979.
5. Berne, E., *Structure and Dynamics of Organizations and Groups,* Grove Press, New York, 1966.
6. Fiorelli, J. A., "Some Considerations in the Selection of Individuals for Team Problem Solving Efforts," *Proceedings, Society of American Value Engineers* **17,** 132–136, 1982.
7. Parker, G. E., "Applying Understanding of Individual Behavior and Team Dynamics with the Value Engineering Process," *Proceedings, Society of American Value Engineers* **24,** 224–229, June 1989.
8. U.S. Department of Defense, *Principles and Application of Value Engineering,* Vol. 1, U.S. Govt. Printing Office, Washington, D.C., 1968.
9. Drozdal, S. C., "Criteria for the Selection of Value Engineering Projects," *Proceedings, Society of American Value Engineers* **13,** 102–106, May 1978.
10. Parks, R. J., "Project Selection—Key to Success," *Proceedings, Society of American Value Engineers* **21,** 71–74, 1986.
11. Fowler, T. C., *Value Analysis in Design,* Van Nostrand Reinhold, New York, 1990.
12. Barker, J. A., *Discovering the Future: The Business of Paradigms,* Video, Charthouse Learning Corp., 221 River Ridge Circle, Burnsville, Minn. 55337, 1984.
13. Welter, T. R., "How to Build and Operate a Product Design Team," *Industry Week,* pp. 35–50, April 1990.

BIBLIOGRAPHY

1. Bachli, L., "Cooperative Value Engineering Team Building," *Proceedings, Society of American Value Engineers* **19,** 106–109, May 1984.
2. Barlow, C. M., "Organizing for Acceptance—Value Management as a Corporate Resource," *Proceedings, Society of American Value Engineers* **13,** 283–289, May 1979.
3. Boothe, W., *Developing Teamwork,* Golle and Holmes, Minneapolis, Minn., July 1974.
4. Fordyce, J. K., and Weil, R., *Managing with People,* Addison-Wesley, Reading, Mass., 1971.
5. French, W. L., and Bell, C. H., *Organization Development,* Prentice-Hall, Englewood Cliffs, N.J., 1990.
6. Gomolak, G. J., "Team Building—Who Said It Would Be Easy?" *Proceedings, Miles Value Foundation* **1,** 125–132, 1988.

7. Hall-Quinlivan, D., and Renner, P., *In Search of Solutions,* Training Associates, Ltd., Vancouver, B.C., Canada, 1990.
8. Kiefer, C. F., and Stroh, P., "A New Paradigm for Developing Organizations," in *Transforming Work* (J. D. Adams, Ed.), Miles River Press, Alexandria, Va., 1984, Chapter 11.
9. Marcon, J., "The Optimum Application of the Value Disciplines—Integrated Value Management," *Proceedings, Society of American Value Engineers* **6,** 161–173, May 1971.
10. Marsh, S., Moran, J. W., Nakui, S., and Hoffherr, G., *Facilitating and Training in Quality Function Deployment,* GOAL/QPC, Methuen, Mass., 1991.
11. Mudge, A. E., *Value Engineering: A Systematic Approach,* McGraw-Hill, New York, 1971.
12. Port, O., Schiller, Z., and King, R. W., "A Smarter Way to Manufacture," *Business Week,* pp. 110–117, April 30, 1990.
13. Rivet, G. J. C., "The Value Engineering Instructor," *Proceedings, Society of American Value Engineers* **11,** 213–219, April 1966.
14. Schein, E. H., *Process Consultation: Its Role in Organization Development,* Addison-Wesley, Reading, Mass., 1969.
15. Schein, E. H., *Process Consultation,* Vol. II, *Lessons for Managers and Consultants,* Addison-Wesley, Reading, Mass., 1987.
16. Scholtes, P. R., *The Team Handbook,* Joiner Associates, Madison, Wisc., 1988.
17. Sehr, E. K., "How to Improve the Efficiency of Value Engineering (VE)," *Proceedings, Society of American Value Engineers* **17,** 93–97, May 1982.

CHAPTER 9

VOICE OF THE CUSTOMER (VOC)

INTRODUCTION

The backbone of the QFD and COPC processes is the voice of the customer (VOC). It is also the basis for product design. It is what started the QFD movement in Japan and the rest of the world. A poor job defining the VOC up front will create problems throughout the rest of the product life cycle, including poor design, overdesign, value mismatch, inflated unit manufacturing costs, excess redesign cycles, problems in the field, and the list goes on!

VOC can also be elusive. How do you know when you really have it? What is the VOC? How many different types are there? When do you use which type? Where and when do you start to collect it? Who should be involved in collecting and structuring the data?

Just as difficult to define is the customer. Who is the customer? Is there more than one customer? Which one(s) should we use to collect VOC?

To make matters more complex, the VOC changes with time. How does the importance of customer verbatims change with time? Can we track the direction of change? Can we estimate the magnitude of change?

VOC is two things. It is the raw data verbatims collected from the customer and it is the team's processed understanding of that raw data. Once structured and interpreted, the processed understanding of the VOC is deployed through the QFD process into the product design community and the company commercialization process.

VOC may also be viewed as a process that involves obtaining, structuring, prioritizing, and measuring customer needs. As such the VOC process itself is separate from the QFD process. The output from the VOC process becomes the input to the QFD process. That is, VOC is the independent process and QFD is the dependent process.

So, keeping this context in mind, let us address all of these concerns one topic at a time.

FUNCTIONS, FEATURES, AND VOC

When designing a product/service, it is helpful to think in terms of a customer-product system. The customer-need product-fulfillment system works as any other. All the elements are linked together.

For example, the customer (user) has a task that he or she wants to perform (see Figure 9.1). This task, in turn, is accomplished through the basic function of the product. That is, the basic function (the reason for existence of the product) is to allow the customer to accomplish a specific task. The basic function is accomplished through a series of operational functions, which we can group into work functions (the basic nuts and bolts), sell functions (performance attributes that differentiate us from other products and attract the user), and perk functions (those that excite the user and make the product more attractive above the sell functions). Product features fulfill cus-

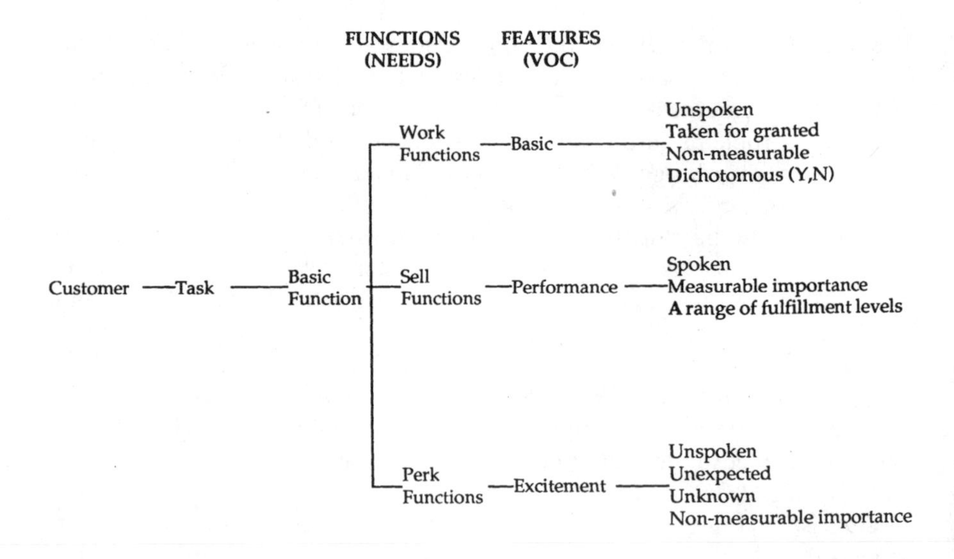

Figure 9.1 Customer-function-feature system.

tomer needs for each function. They are the physical fulfillment of a need. These are the traditional customer needs and are often labeled as VOC. Basic needs features pertain to work functions and are the company's ticket to play the game. I expect my car to start. I expect my car to be safe. They are expected by the customer and the manufacturer gets no credit if they are there. Conversely, you can lose big if they are not there. The best you can do is break even. Basic needs are unspoken and nonmeasurable; they are either satisfied or they aren't.

Performance features pertain to the sell functions. They are spoken needs. They can be measured for importance as well as for a range of fulfillment levels. The customers tell us or answer our question about what they want. I want a car that goes from 0 to 60 in 12 seconds; I want my car to get 24 mpg. This is where it is best to bring in Business Research for assistance. The manufacturer who consistently supplies these features the best will get the sales. These are differentiating needs and differentiating features. This is where marketing comes into play. This is where advertising spends most of its time. This is the battlefield.

Excitement features are future oriented and usually high-tech. They pertain to the perk functions. These are features that the customer is not aware of. They come from the R&D shelf. Air conditioning in a car in the middle 1950s is a good example. Today we have ABS brakes, air bags, voice-activated phones. They are exciting but are not quite to the point of being standard equipment. In the future some manufacturer will introduce a satellite autonavigation system. High-tech but still on the development shelf. These needs are unspoken and unexpected because the customer does not know they exist other than in science-type magazines. They are too futuristic to measure for importance.

The perk functions contain the signals for the future. New features begin here and migrate toward being basic function features. For example, in a few years ABS brakes will be standard items on certain lines of cars. In the 1940s, cars did not have turn signals; today they are taken for granted as standard equipment. The tools of technological forecasting (Chapter 6) can be used to look into the future to see what is on the horizon for perk functions. Products will track technology. The company that goes public first with excitement features sets the trend, and in many cases sets the international standards, such as in cassette tape and CD formats.

Figure 9.2 illustrates the customer-function-feature system for a three-hole punch. The customer task is to organize papers. It is not to punch holes! The basic function of the punch is register holes. In order to register holes the fundamental functions of locate sheet and pierce sheet must be performed. These are assumed by the user, as are some basic features like, "easy to insert paper" and "paper stays in place once I insert it." We also have some sell functions such as being able to relocate the punch assembly to accommodate other paper-hole formats. Collect waste is a convenience function, especially if our design prevents holes from falling all over the office. Finally, a perk function

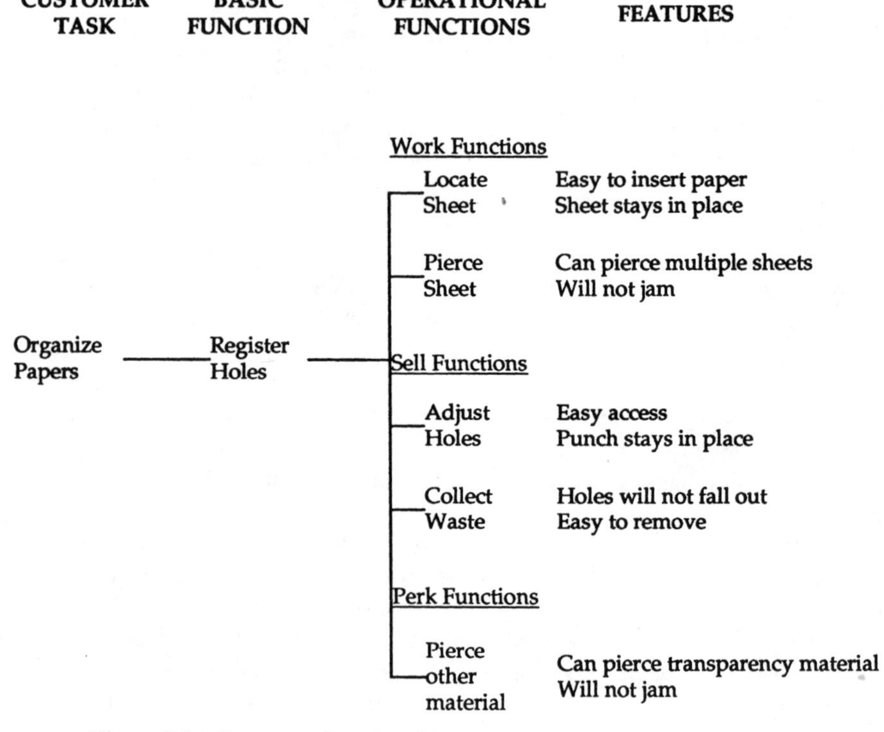

Figure 9.2 Customer-function-feature system for a three-hole punch.

might be: pierce other material such as overhead transparency material, which is difficult to punch without jamming. The company who makes a three-hole punch that will punch this tough material may just have an edge.

If we view supplying a consumer product as a system (Figure 9.1), we can see that if we change a feature or a function, or if the customer changes a need or even the task, the system reacts. Things change. Companies must be alert to these changes. Their antennae must always be raised to catch new signals. Many of these signals, but not all, are contained in the VOC. Technological forecasting, discussed earlier, is used to capture signals of technological change.

VOC is where the action is. The company that does a good job capturing the VOC and delivering the product the fastest is the one who is going to win. Notice that quickness in getting to market is a key issue. Incorporating VOC into the product will not, in itself, guarantee success.

Where does quality fit into the picture? I am afraid that today quality is quickly becoming a hygiene factor. Either you have it or you don't and if you have it, it must be the best. As we approach the 21st century, quality is being taken more for granted. Quality will not be a differentiator because all competitive companies will have it. Timing and VOC will be where the battle will be won.

So, let's talk more about VOC.

SUPPLY AND DEMAND

Figure 9.3 is a supply-demand function diagram[1] that describes broad functions involved in obtaining and using VOC data. The demand side of the diagram, shown on the left-hand side of the center scope line, answers why the

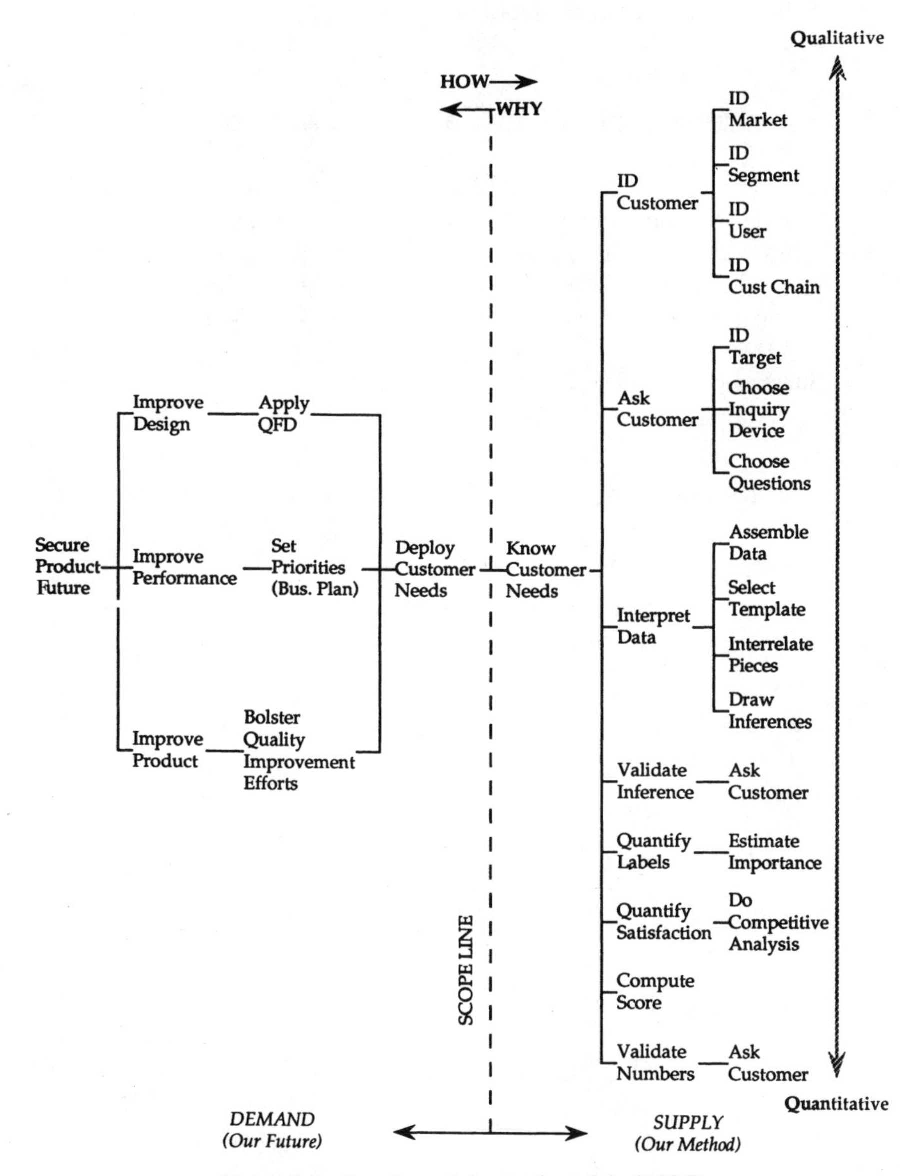

Figure 9.3 Supply and demand model of VOC.

VOC is obtained. It is company and future oriented. Methods for meeting the demand are shown on the right-hand side of the diagram and show how the VOC can be obtained. This side is methods oriented and deals basically with the present. Qualitative methods are shown at the upper right of the diagram, indicating that they are the first tools to be used. Notice that the methods become more quantitative as one progresses down the list of functions. The entire diagram is based on how-why logic; how is asked from the left to right side of the diagram, and why is asked in the opposite right to left direction. The answers should check in both directions.

VOC COLLECTION TECHNIQUES—WHICH? WHEN? HOW?

Deciding how and when to collect the VOC is half the effort. Many times this can be decided by comparing the different collection techniques. It is wise to work closely with the business research department or equivalent of your company. After all, this is their line of work. It is important to remember, don't do it alone! The most common collection methodologies are:

1. Surveys, mail, phone, comment cards.
2. Interviews, group or individual, phone.
3. Focus groups.
4. Location studies (like industrial engineering surveys).
5. Direct observation. Visitation to site of usage.
6. Internal brainstorming.
7. Commercially prepared stock reports.
8. Conjoint analysis.
9. Panels.
10. Electronic databases and searches.
11. Service calls.
12. 1-800-hotlines

Sources of VOC Data	Type
1. Internal (marketing, sales, customer service)	Recorded
2. Customer direct	Observational
3. Trade shows	Observational
4. Sales calls	Observational
5. Service/repair calls	Observational/recorded
6. Literature (trade and consumer)	Recorded
7. Complaints/warranty records	Recorded
8. Visitations	Observational (subjective)
9. Electronic databases and searches	Recorded

Answers to the following set of questions can help you determine the right collection technique for you and your project.

1. What is the collection technology? How does it work?
2. When is the best time to use it?
3. How long will it take?
4. What will it cost?
5. What are the advantages?
6. What are the disadvantages?
7. What level of detail do you get? System? Subsystem? Piece/part? Or something else?
8. What are the prerequisites for using this collection method?
9. What are the assumptions? How much bias is implied?
10. What is the output? What do you get? A report? A computer disk full of numbers?
11. What is the lapse time to output?
12. What is the format of the output? Labels only? Numbers? Levels? All the above?
13. Do you know what questions you want to ask? How were they derived? Did you use the QFD matrix to assist in generating the questions?

There is no one VOC collection technique that will give you everything you need. Use several techniques in concert to help you cover all the bases you wish to cover. Also, different VOC techniques support each other. If several different collection techniques come up with the same answer or point in the same direction, this lends more credibility to the study. It is highly unlikely that they are all wrong.

QUALITATIVE AND QUANTITATIVE VOC

There are two types of VOC data, qualitative and quantitative, or linguistic and numerical. The qualitative VOC data is in the form of labels. It is subjective and exploratory and tends to be open-ended. It can be both divergent and convergent. It provides structure, surfaces relationships, and forms a database for further focus.

Quantitative VOC, on the other hand, is numbers oriented and tends to be more objective and specific. The numbers show magnitude and assist us in our dialogue in being more convergent. They provide metrics for focus within a qualitative database. This type of data is more business research oriented. Numbers show importance of and the strength of qualitative relationships.

Which of the two types do you use and when? First, it is important to establish a database using qualitative VOC. We have to get the labels on things before we can measure any thing. Figure 9.4 depicts the shifting balance between qualitative and quantitative VOC data over the product life cycle from vision to sales.

The slope of the line between qualitative and quantitative VOC is determined by how well we understand the VOC. That is, the more we understand the VOC, the faster we change the balance between qualitative and quantitative VOC over time. This is more commonly known as experience and learning. Like other things the VOC also has a learning curve.

Next we apply the most basic numerical measures (subjective estimates) to determine the hierarchy of importance across the labels of the qualitative database. The combined qualitative and quantitative data will highlight what we know and what we don't know about the VOC. At this point it is equivalent to completing the rows of the HOQ or the left side of a COPC matrix. Those items about which we are uncertain, about the label, about its numerical measure, or both, are highlighted by a yellow marker. Once the VOC database is finished we can easily and quickly scan the entire matrix for unknown and missing VOC data. These highlighted, questionable VOC data become the input we use to plan a more organized, specific, and focused inquiry. At this point it is helpful to call in business research and/or marketing for help in designing surveys and focus groups. The activity now resembles that which is used in more traditional market research, but we now have a firm foundation from which to build better research.

WHEN DO YOU START?

The majority of QFD teams believe that in order to do QFD you have to first run surveys, focus groups, interviews, and so on, to collect the VOC before doing the HOQ or COPC. My question at this point is: How do you know this early in the project what questions to ask? Sometimes market, segment, and user have not yet been well defined. There is a risk at this point that such VOC collection will be a shot in the dark. It too easily can produce a database biased by strong-willed individuals covering up what they don't know.

My suggestion and my mode of operation is to take all existing internal VOC data and, using affinity and tree diagrams, construct the rows of the HOQ first! Rate the customer verbatims for importance, do a competitive analysis, and rate market leverage. This will give order and structure to what we already have. It will tell us what we know and what we don't know, as just discussed in the previous section. Knowing and structuring what we don't know now puts us in a position to formulate more and better questions, to design focus groups, interviews, and other VOC data collection methods. New data would be entered into the blank spaces and replace questionable labels and numbers in the preliminary HOQ.

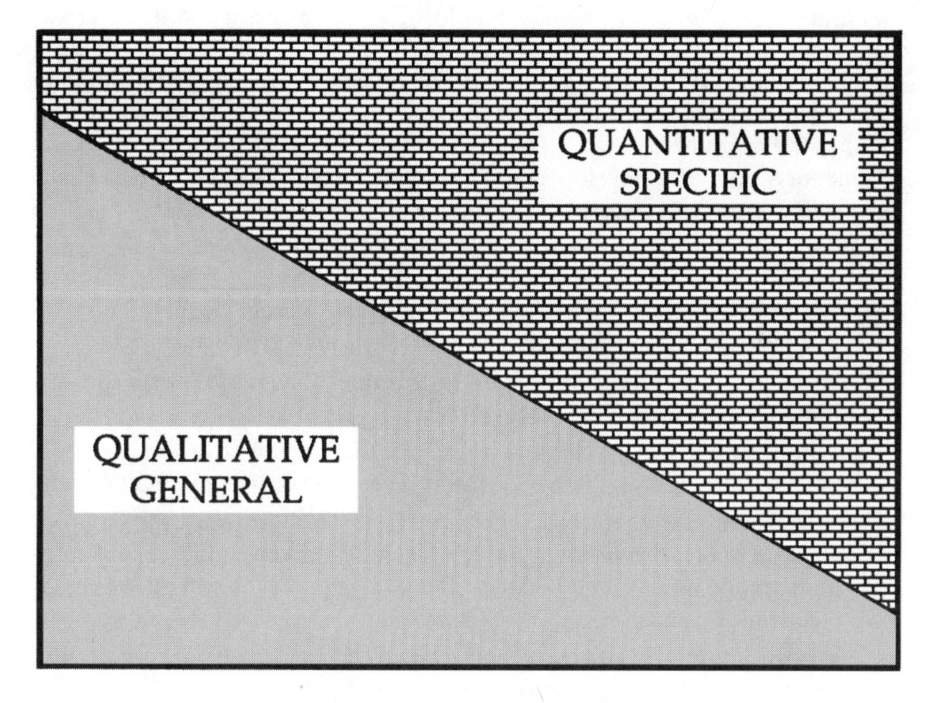

Figure 9.4 Ratio of qualitative and quantitative VOC.

Remember the HOQ, in particular the VOC data, is a living document that should continually be updated. Therefore, we should be obtaining VOC all the time from trade shows, sales calls, customer service, and so on.

WHO IS THE CUSTOMER?

Before we attempt to collect the VOC, we must first establish who the customer is. Many times this is not an easy task. The difficulty arises from the fact that there is usually a chain of customers. These customers can be internal and external. Each one in the chain has his or her own set of needs (VOCs). So, for whom are we designing the product? Who is affected by our product? Is it the final user? The purchaser? A person who recommends? All the above? One might argue that we design for everyone in the customer chain, but doing so increases the chances for an overdesigned, overpriced product. Here is where marketing must be consulted because they usually set the target, in conjunction with the business plan, for country, market segment, and user. This is another example why it is necessary to use a cross-functional team. Marketing and/or business research must be participating members of the QFD team! Getting started questions are covered in detail in Chapters 7 and 8. They are some of the first questions to ask in the effort to identify the customer.

One method I have developed and used with considerable success is what I have called a customer morphology matrix. The matrix consolidates possible answers to the following questions (see example in Figure 9.5):

1. Where is the product used? All possible areas of usage begin to illustrate the various market segments. List all areas where product is used.
2. Who uses the product? Answers to this question start to form the customer/user chain. List all users.
3. Who makes the buy decision? These are the individuals who have a significant influence on the buy decision. They may or may not be the user!
4. What method is used (to handle product)? This illustrates the environment in which the product is used, such as manual mode versus automatic mode, mainframe versus personal computer, existing light versus electronic flash.
5. Whose product is currently being used? What brand(s) are currently being used within the answers to Question 4?

For all of the options under each question above, it is revealing to list the number of units sold and convert units to a percentage. Likewise, list dollar volume plus dollar volume percentage. For other questions (like: Who uses? Who makes the buy decision? What method is used? What brand is used?) an estimate of the percent occurrence for each item is entered. All percentage figures by question must sum horizontally to 100 percent, as in the example.

Now that we have labels and percent occurrence of items for each question, we can search for patterns. We can develop Pareto distributions. For example, we could circle all of the highest percentages and this would give us a vivid picture where the activity is. Likewise, we could draw a corral around the lower percentages, which would represent possible areas for market opportunity and market development. The corral drawn in our example (Figure 9.5) indicates where sales are coming from, who is the most active user, who generally makes the buy decision, the predominant method of handling the product, and which manufacturers have the largest volume according to method.

We now have an educated boundary around who the customer is. Attempts at documenting the VOC are best started after the customer or customer chain has been documented. Not documenting the customer results in repeated false starts and lost time defining the VOC on the wrong customer.

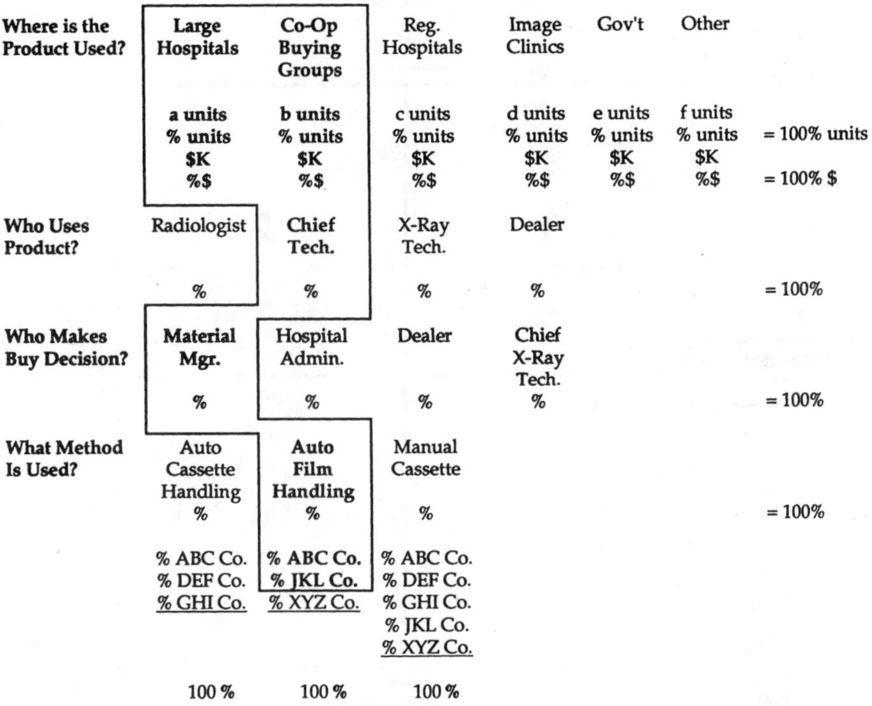

Where is the Product Used?	**Large Hospitals**	**Co-Op Buying Groups**	Reg. Hospitals	Image Clinics	Gov't	Other	
	a units	**b units**	c units	d units	e units	f units	
	% units	**% units**	% units	% units	% units	% units	= 100% units
	$K	**$K**	$K	$K	$K	$K	
	%$	**%$**	%$	%$	%$	%$	= 100% $
Who Uses Product?	Radiologist	**Chief Tech.**	X-Ray Tech.	Dealer			
	%	**%**	%	%			= 100%
Who Makes Buy Decision?	**Material Mgr.**	Hospital Admin.	Dealer	Chief X-Ray Tech.			
	%	%	%	%			= 100%
What Method Is Used?	Auto Cassette Handling	**Auto Film Handling**	Manual Cassette				
	%	**%**	%				= 100%
	% ABC Co. % DEF Co. % GHI Co.	% ABC Co. % JKL Co. % XYZ Co.	% ABC Co. % DEF Co. % GHI Co. % JKL Co. % XYZ Co.				
	100 %	100 %	100 %				

Figure 9.5 Customer morphology. The bounded entries represent the playing field and the players we wish to engage in competition.

OTHER TOOLS FOR STRUCTURING VOC

Another useful tool to expand VOC is the voice of the customer table (VOCT) developed by GOAL/QPC.[2] The VOCT is a method to gain more in-depth understanding of needs by expanding current needs through the use of the 5W1H questions (what, where, when, why, who, and how). Expansion would not be performed on all of the VOC verbatims. Generally the most important ones are selected for further study. GOAL/QPC uses the VOCT before doing an affinity diagram and the HOQ. The VOCT is a tool to further explore what customer needs may be in the future in order to derive more needs of the exciting quality variety. The VOCT can be used internally within the QFD team and may or may not be used directly with the customer.

A nice companion to the HOQ is the Customer Window℠, developed by Arbor, Inc.[3] This is yet another tool to segment customer needs. It helps give another perspective in our effort to locate customer needs items to work on during our effort to improve product. Arbor's Customer Window℠ (Figure 9.6) is part of their larger comprehensive process of obtaining and classifying customer needs to achieve customer focused quality improvement. I have used the Window in strictly a qualitative capacity to locate those needs the customer

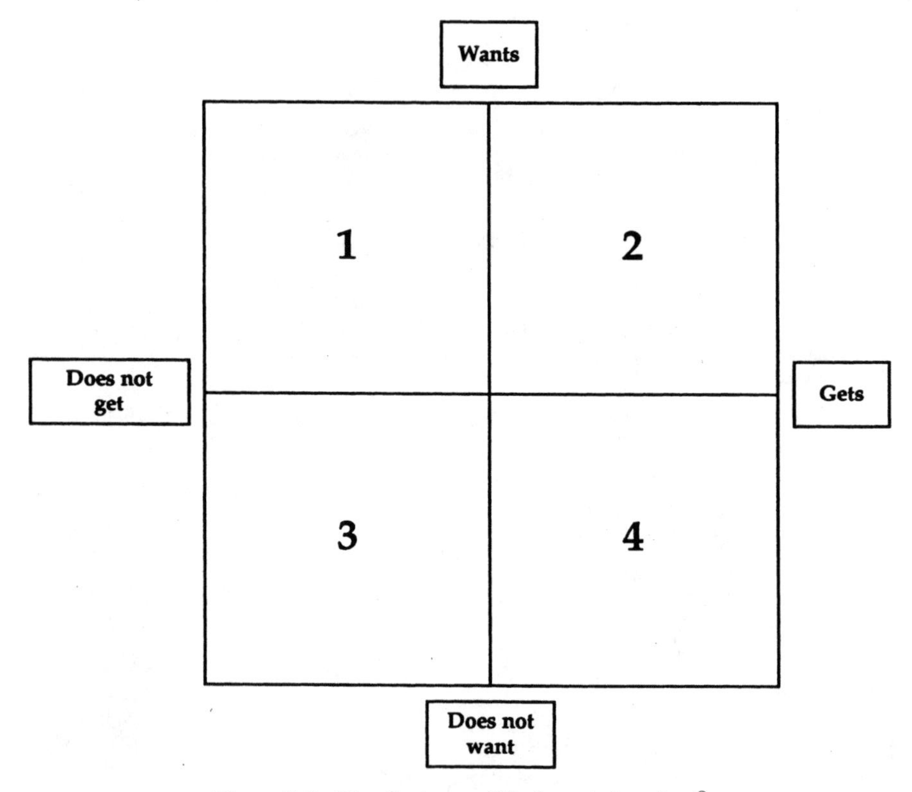

Figure 9.6 The Customer Window, Arbor, Inc.©

wants but does not get. The mechanics of the Customer Window℠ are simple. It is a grid where the x-axis represents the degree of fulfillment of a need from "does not get" to "gets." The y-axis represents the desire of the customer for the need from "does not want" to "wants." Plotting the customer verbatims on this grid gives revealing insight as to how your product is doing. Those needs residing in quadrant 1 are targets for improvement and expansion of product design. Those needs residing in quadrant 4 are candidates for elimination and cost reduction. The money saved by eliminating items in quadrant 4 can be used to develop and include items in quadrant 1.

The grid plots can be used as a double check on the HOQ. For example, items in quadrant 1 would be expected to represent a blank row in the correlation matrix in the HOQ. Items in quadrant 4 are likely to represent a blank column in the HOQ.

LISTEN CAREFULLY! TRANSLATING THE VOC

One has to listen carefully when talking to the customer about what he or she wants in your product. This is also true when compiling survey comments.

Many times customers will offer you solutions to some of their unsolved problems. Sometimes they may even state their needs in terms of problems. Obviously these problems have been around for some time and our product has not eliminated them. The customer in many of these cases is aggravated and has decided to come up with a solution to pass on to you because this customer figures you cannot solve the problem. Consequently we are given solutions disguised as needs!

In these cases further, more penetrating questions must be asked to seek and clarify the real need. Questions can be asked for clarification such as:

1. What is the unmet need that this solution is satisfying?
2. What is the unsolved problem this solution is satisfying?
3. Why is this a problem?

Let me illustrate with an example. I purchased a minivan and a company representative is interviewing me about my experience and needs for the car:

Customer States "What": "I want splash guards as standard equipment on my van!"

QFD Interviewer: "What is the need that this solution would satisfy?"

Customer: "Splash guards will prevent stone chipping and sand blasting behind the fender wells and rocker panels."

QFD Interviewer: "Why is this a problem?"

Customer: "Chips in the paint in and around these areas can quickly turn to rust, especially here in my home town where we use lots of salt on the streets in the winter."

What is the real customer need?

Based on the above dialogue and appropriate questions from the interviewer, we can quickly see that the real need is to prevent rust around fender wells. Splash guards are just one of many ways, in this case the customer's, to alleviate this problem. The customer, with an extra $20, decided to address the problem himself. We can also see that rusty fender wells would also make an excellent QFD project.

Even clearly stated needs can be explored further with the right questions. The following questions would take a customer well beyond his or her original comments:

1. Why is this a benefit?
2. You want ______so you can______? Why?
3. Why do you want that?
4. What do you really want?

Such questions may uncover new hidden needs the customer didn't think about. Sometimes stated needs can be the object of unfulfilled fantasies. These fantasies are in the realm of exciting quality in Kano's model.[4] They are also the principle behind the VOCT discussed earlier.

GLOSSARY

A glossary of terms for customer verbatims is very helpful and can save the team a considerable amount of time over the life of the QFD team. Many hours of communication can be lost because team members had their own interpretation of a customer need. Building a glossary at the outset improves communication among team members. The dialogue that must take place to build the glossary is also an education process where team members learn things they never knew before. This subject is also briefly discussed in Chapter 4. A good set of terms is needed both for the VOC and for the product technical requirements in the HOQ. The correlation and ranking processes require a good understanding of the items. Good definitions help the correlation and ranking process by promoting consistency of interpretation, which in turn helps promote consensus.

EXTENDING VOC INTO THE FUTURE

Whenever VOC is determined and documented for QFD applications, it is generally done in the present tense. That is, we derive the labels for customer needs and then rate their importance as they exist today. It is a static slice of time. Since many QFD teams are involved with the design of future products, the team members are quick to realize that current importance of customer needs may not be the same in the future. Importance can increase or decrease or perhaps the need may disappear altogether. As former needs disappear, new ones may come into existence. In order to remain competitive, it is necessary to reevaluate customer needs on a periodic basis. To do this requires a process and a structure to peer into the future.

A trend matrix (Figure 9.7) can be constructed to register team perception of importance time phased into the future. Input regarding importance should come from the best sources. This may require contacting individuals outside the team, from professional societies, academia, customers, information banks, electronic data search firms, and so on, to develop a trend for each customer need. Also it is necessary to be alert for new needs that don't exist today. A monitoring file, as discussed in Chapter 6, is appropriate for starting a VOC database. The trend matrix is constructed by first listing the customer needs, usually from the HOQ or COPC matrix, and then recording how important those needs are today for the particular market, segment, and user. Generally a 1 to 5 scale is used to express current importance. Descriptors are derived for each numerical anchor point of the scale, as was done in Chapters 2, 3, and 4.

VOC Needs	Current Importance	Future Importance	
		5 yrs.	10 yrs.
Productivity	3	↑	↑
Reliability	5	→	→
Maintenance	4	→	→
Cost/page	4	↑	→
Disposability	4	↑	→
"Green"-ness	3	↑	↑
Media interface	3	↑	↑
Standard interface	3	↓	0
Future	0	↑	↑

↑ **Increasing importance**

→ **Status quo**

↓ **Decreasing importance**

0 **Need is/will not be a factor**

Figure 9.7 VOC trend matrix.

To indicate the trend in importance in those needs over time one can use a series of arrows. An arrow pointing up indicates the current need will increase in importance; a horizontal arrow indicates the importance will remain as is; an arrow pointing down indicates that the importance will decrease; a zero indicates the need will disappear.

I prefer to use arrows signifying direction because it is difficult to indicate future need with a rating scale. All that is important is the direction in which the need will be changing. That is, we want to know the direction of the arrow, not its length.

One useful method for collecting VOC information about the future is the Delphi method using a Delphi questionnaire.[5] This method was developed by the Rand Corporation in the 1950s and has many variations based on the original model. Figure 9.8 is an adaptation that can be tailored to collecting VOC

PART A What is likelihood that:	5 Yrs: Short Term	10 Yrs: Long Term
1. Productivity will be at least 100 units/hr.?	90	100
2. Recyclable consumables / containers will be federal law?	85	100
3. etc.		

PART B Using arrows, ↑ increasing, ↓ decreasing, → no change, estimate the change in importance of each of the following needs (If you believe the need will disappear, use a zero):		
1. Productivity	↑	→
2. Reliability	↑	↑
3. Maintenance	↑	→
4. etc.	↓	0

Figure 9.8 Delphi survey form for VOC.

data about the future. Questionnaires would be summarized and the summary data would be entered into the VOC trend matrix in Figure 9.7.

There are numerous ways to conduct a Delphi survey. Two types of questions that have worked well are: (1) What is the likelihood that a particular event will occur by a specific time frame? (2) What is the direction of current trends into the future?

Part A of Figure 9.8 asks respondents the likelihood of an event occurring within five years and ten years. For this example, five years and ten years are arbitrary and would be adjusted according to the subject area studied. A rating guideline should also be included to help respondents estimate probability. For example: highly likely = 90–100%; quite likely = 65–90%; as likely as not =

50%; not very likely = 10–35%, and unlikely = 0–10%. It is important that a category scale is used.

Part B of Figure 9.8 asks respondents to draw arrows in the answer space to represent their perception of the future trend of particular events. Arrow category descriptions are listed on the questionnaire for trending guidelines.

The original Delphi model used yet another questioning method. The original questioning technique asked people to estimate how many years they thought it would take until a particular event occurred. For example, how many years will it be until all new U.S. cars after 1995 have ABS brakes as standard equipment? This type of question can also be used for VOC. Whichever variation is used, the wording of the questions is very critical to be sure you elicit what you want.[6] The choice is yours.

Knowing the future direction of customer needs is helpful in selecting what technologies to use for future products. For example, if the COPC matrix is kept up to date regarding technology, future generations of products can be speculated upon, based on selection of latest technology and estimated importance of needs for the approximate time period of product introduction. In this respect the COPC matrix becomes a technology library for future product development.

Another Delphi variation is an interactive grid mapping process. The process works on the same principle except that the various customer needs are written on cards (PostIts). At the interview, the interviewee is asked to place the cards on a grid (Figure 9.9) where the y-axis represents time and the x-axis represents impact. If the interviewee develops other needs that are not in the starter deck of cards, new cards are written and are placed on the grid. The interviewee may also be asked to estimate the probability of occurrence on the card. The resulting grid is a map of when certain needs will occur, their probability of occurring, and the impact if they do occur.

MONITORING THE VOC

In addition to extending the VOC into the future, there are areas of uncertainty uncovered by the QFD process that should be monitored into the future. Monitoring, as previously discussed in Chapter 6, is used to track signals and events in order to assess change. These signals could have a profound impact on the company or project if they ever reach a critical mass. The QFD/VOC process helps us discover what signals to monitor. Monitoring has been discussed in more detail in Chapter 6.

Monitoring and Delphi data also provide input to constructive scenarios about the market and customer. Scenarios also are discussed in Chapter 6.

HOW MUCH DETAIL?

It is difficult to decide how much detail to go into until one starts collecting the data and verbatims. Usually, VOC is collected at the third level of a tree dia-

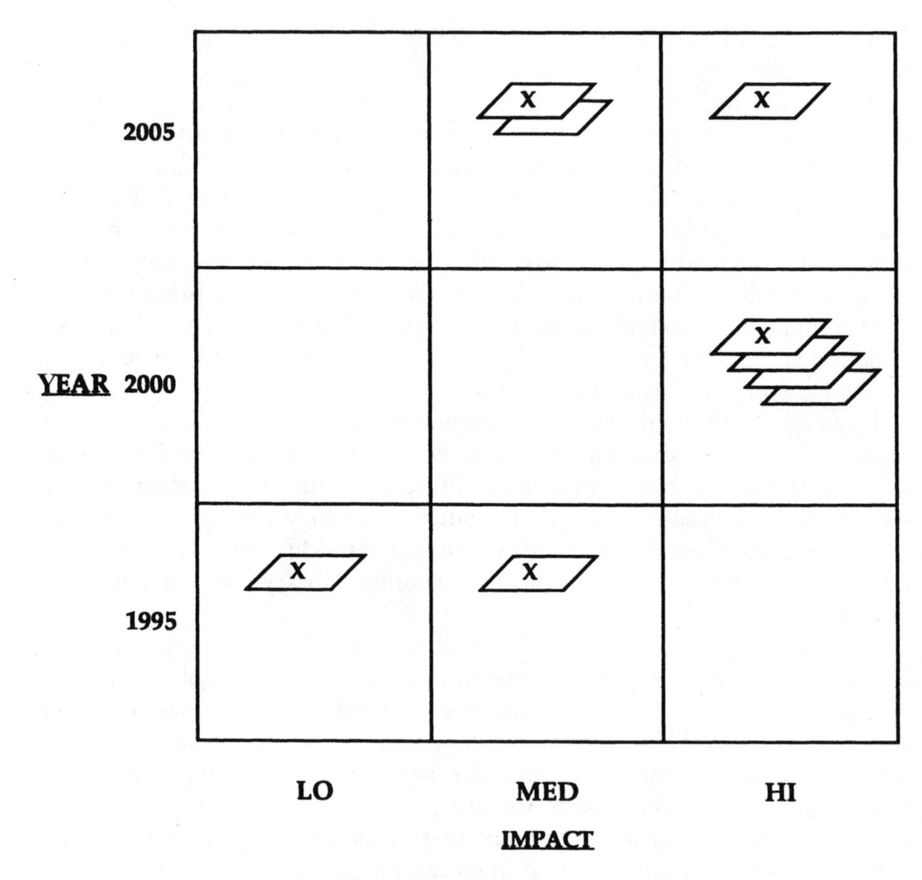

Figure 9.9 Interactive delphi grid.

gram to be sure that everything is covered. It is easier and less costly to collect too much at first than to go back out a second time to collect more.

At what level is the team comfortable? If collecting and categorizing at the third level creates a 2000 cell matrix, many people will be discouraged from finishing the matrix. QFD should not be a long drawn-out horror show! The initial QFD (HOQ) should be at a macro total system level of detail in order to find those areas that should be broken out into smaller QFD's at a micro subsystem level of detail.

To do this, start at the VOC second level of detail in the affinity diagram and the tree diagram to build the initial HOQ. The most important areas will surface at this level. Each of these areas could be treated as a smaller individual QFD at the third level of detail. The team could be broken into smaller units, each working on its own micro HOQ representing its members' area of expertise.

VOC level of detail will also be helpful in collecting and clustering the product technical requirements.

Many QFD projects have died because the matrices were too large. They collapsed under their own weight and the people became turned off.

TRANSIENCE OF THE VOC

The VOC is transient. It changes with time. Labels change and importance ratings change. This can create problems in product design. In design, we cannot aim at a moving VOC target because doing so will increase the number of redesign cycles and quickly and significantly drive up cost. We must pick a static slice in time in a dynamic VOC world. We have to pick a date where design is to be frozen. All changes to product beyond this date will be included in the next design model at a later date. This is the Japanese approach of introducing many models in rapid succession, each with small incremental changes/additions. Each model takes advantage of all prior accumulated experience to date. In the United States we usually feel we have to hit a home run, which prolongs the development cycle as we try to get it right. In the final analysis we tend to miss the target; too much too late.

The VOC, as well as the QFD/COPC process in general, is a dynamic process that pulses forward and loops back as needed. The process should loop future signals back to the shifting present (see Figure 9.10). The QFD and COPC matrices then are living documents that are updated over time. Keeping the matrix data updated makes it much easier to design the next product model.

HOW MANY VOCS?

What do we mean by VOC? Is there more than one voice? I believe there are three voices. The voice of the customer (VOC), the voice of the company

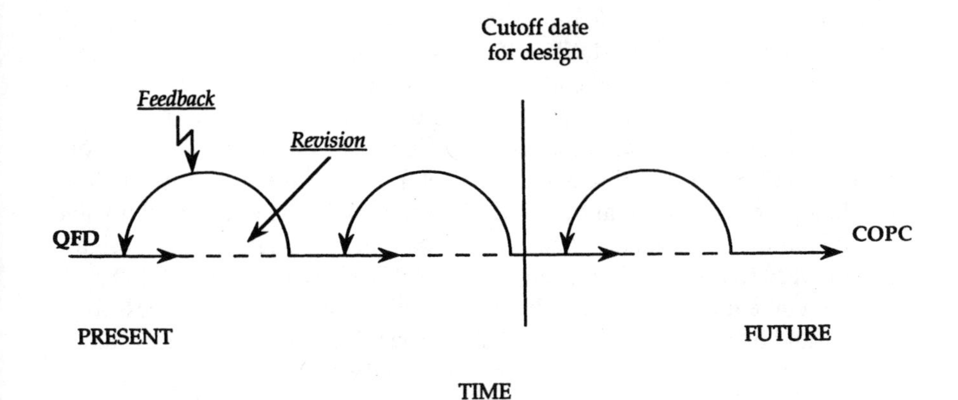

Figure 9.10 Pulse forward-loop back VOC model.

(VOCo), and the voice of the engineer (VOE). All of these voices have needs and we cannot attempt to satisfy one voice at the sacrifice of any of the others. They all have to be considered. How do we integrate all three? This is the very situation that gave birth to customer-oriented product concepting (COPC), discussed in Chapter 4.

Notice in Figure 4.1, depicting a COPC matrix, there is a column titled "Customer requirements," which consists of "Features" and "Importance"; this collection of data represents the VOC. The next column, labeled "Competitive analysis," may also be included in VOC.

The column titled "Manufacturing criteria" represents another voice, the voice of the company. Finally the numerous columns labeled "Technologies" represent the voice of the engineer (VOE) or voice of the designer (VOD).

The construction and utilization of the COPC matrix provides the structure for the team to consider all voices simultaneously. If the rules of COPC are adhered to, all voices must be given equal air time.

Remember, in Chapter 8 we discussed answering two basic starting questions before initiating a COPC or QFD. Those questions are: (1) What are we trying to do for the customer? (2) What are we trying to do for the company (ourselves)? The three voices are indentured under these two questions. The VOC obviously helps to answer question number one, what are we trying to do for the customer. The VOCo and VOE are directed toward answering the second question, what are we trying to do for the company. To reiterate, our attempt to answer and document one of the two questions cannot be at the sacrifice of doing a poor job on the other. There must be a good balance in satisfying needs between customer and company.

Actually, when we think about all these different voices, it is hard to do so without implying some sort of stratification. All of these voices enter the product development process in different layers. These layers can represent the levels of perspective (strategic, tactical, and operational), and sometimes levels within the company hierarchy. Figure 9.11 is an illustration of the stratification.

The top part of the triangle represents the strategic level and must reflect the voice of management (VOM). The 12 questions discussed in Chapters 7 and 8 are a useful mechanism to collect VOM. Matrix P1 in the PQFD model (Chapter 5, Figure 5.2) can also serve this purpose. The middle portion represents the tactical level and integrates the VOC and the VOCo and, to some extent, the VOE. COPC, discussed in Chapter 4, is the methodology of choice to build a product concept at this tactical level. Finally, the lower level involves the VOE at an operational perspective to complete a final product design to carry forward to the end. This operational level is where traditional QFD really begins to come into its own. It is the design and problem solving process to put together and drive design down to the shop floor and statistical process control. Answers to the engineer's or designer's VOE make this possible.

For the entire product development process to work, all of the voices from all levels must be compatible, synchronized, and congruent. The chain of

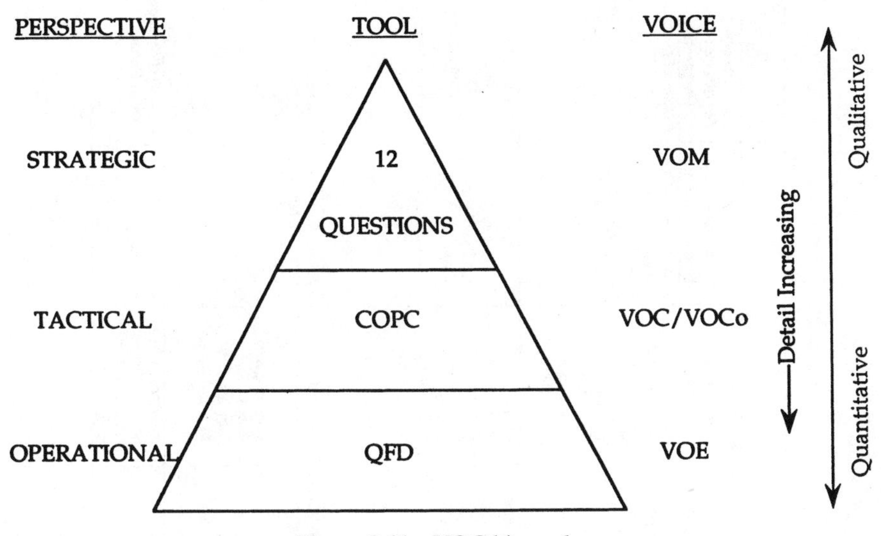

Figure 9.11 VOC hierarchy.

voices forms the communications link and becomes the spinal cord for the company product commercialization or total quality management (TQM) program. The various voices serve as disks in the spinal cord.

Finally, all of these voices must be integrated into a balanced scenario, which I refer to as the collective voice (CV). The quality of this CV scenario is what can make the difference between a great product and an OK product, and, in the worst case, between winning and losing.

To illustrate the interconnectedness of all the elements I developed a matrix, Figure 9.12. The column headings across the top represent the functional steps in the VOC process from the strategic level through the operational level. Down the left-hand side defining the matrix rows are the activities that should take place in each function column. The second row lists all of the tools that can be used at each step of the process. This listing of tools is by no means complete. Row three represents the data sources and documents that are appropriate for each VOC process function. These sources are generic and each company will have its own data sources in addition to those listed here. Finally, the last row lists the products or output of each VOC function.

To simplify the matrix and the VOC process I have used four VOCs. VOM represents the voice of company management, VOMKT represents the voice of marketing, VOC/VOE represents both the voice of the customer and the voice of the engineer, and VOC is the voice of the customer but in a product usage and feedback mode. These voices closely correspond to the levels illustrated in Figure 9.11.

Hopefully the matrix will illustrate the interaction and complexity of the VOC process. It also illustrates what must be in place in order to do a proper QFD or COPC. Too often the contents of Figure 9.12 are assumed and a QFD

VOC PROCESS

FUNCTION	VOM	VOMKT	VOC/VOE					VOC
	Set Global Strategy	Frame Market	Collect VOC Data	Interpret VOC Data	Structure VOC Data	Quantify (Prioritize) Data	Deploy Processed VOC Understanding	Observe/ Feedback Product Performance
ACTIVITY	• Define Market • Develop Business Plan/Strategy • Segment Market • Scout Future • Survey Impacts	• Develop Product Family • Develop Product Strategy	• Select Targets • Select Data Sources • Select Collection Method • Select Personnel • Collect Data • Process Solutions and Unknowns	• Process Customer Solutions and Unknowns • Identify Which Items to Carry Forward • Process Compound Statements	• Show Relationships • Cluster Similarities • Plot	• Select Appropriate Value Measurement Tool • Quantify VOC/VOE Verbatims For Importance	• Incorporate VOC-3 Into HOQ Or COPC • Deploy to Subsequent QFD Matrices As Fit	• Collect Feedback Data • Structure Feedback • Send to Management For Update of VOM, VOMKT
TOOLS	• Strategic Planning • Decision/Risk Analysis • Scenarios • Trend Analysis • Impact Analysis	• Framing • 12 Q's • Is/Is Not • Pareto Distribution	• Personal Customer Contact • Survey • Interview • Focus Group • Direct Observation • Direct Visit • Customer Panel • Customer Council • Product Mentors	• VOC Table • 5-Whys	• Tree Diagram	• Value Measurement • Ranking • Pair Comparison • Scaling • Constant Sum • Scoring Model • Conjoint • Cluster Analysis • Multi-Dimensional Scaling	• COPC • QFD-HOQ • QFD - Higher Matrices	• 1-800-HELP • Industrial Engr. Studies • Customer Visits • Tech. Reps. • Opinion Survey
SOURCES	• VOC • Competitive Intell. • Business Intell. • Customer Contact • Managers	• VOC • Personal Customer Contact • Business Research	• Direct Customer Contact • Complaints • Warranty • 1-800 • Publications • Business Research	• Personal Customer Contact • Collected Verbatims	• Processed VOC/ VOE Verbatims	• Personal Customer Contact • VOC-3 Verbatims • Marketing Best Guess	• Processed VOM, VOMKT, VOC/VOE	• Sales Records • Complaints • Warranty • Direct Observation • Tech. Reps. • Sales People • Dealers • Channel Partners • Repair Centers
PRODUCT	• Business Plan • Technology Strategy	• Product Families • Product Family Strategy	• Unprocessed Customer Verbatims	• Processed Understanding of Verbatims	• Structured Verbatims	• Structured VOC/ VOE Verbatims Quantified For Importance	• HOQ • COPC	• Satisfaction Report • Other Reports

Figure 9.12 VOC process guide.

project gets into trouble quickly. When things don't work out well, QFD usually gets the blame!

PUTTING IT ALL TOGETHER

Figure 9.13 is a flow chart for developing processed VOC. This process is equivalent to deriving VOC/VOE as just discussed. As such, the VOC/VOE process assumes that the strategic plan, the business plan, and the technical and marketing plans are in place.

The first step is to define the scope of the VOC project. VOM and VOMKT from Figure 9.12 are necessary for defining scope. What country, what market,

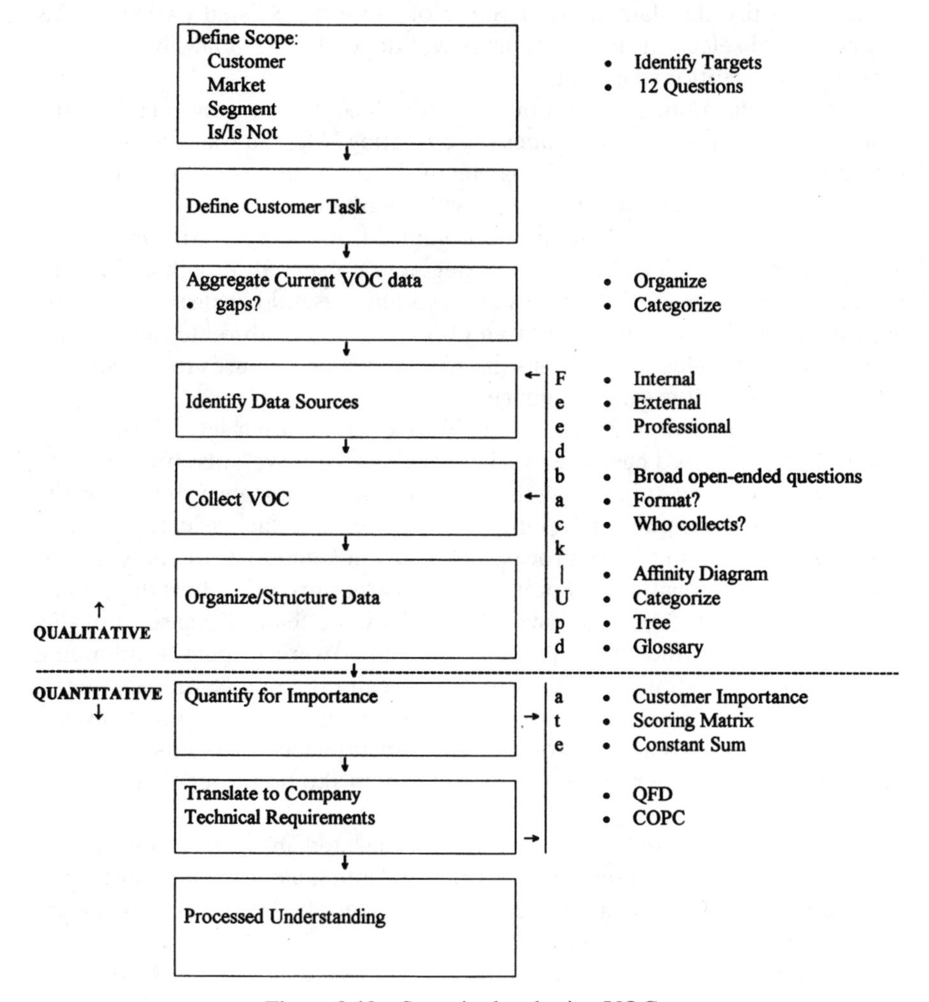

Figure 9.13 Steps in developing VOC.

what segment, and what user are we to learn more about? You cannot construct the VOC unless you know who the customer is. Also important is to determine what the scope of the study is. What is included and what is not included? This will also be affected by market segmentation. However, within segment there may be other particular exclusions. What are they? The 12 questions discussed in Chapters 7 and 8 are useful for this purpose.

The next step is to determine the customer task. Why is the customer buying your product? What does the customer want to accomplish? Tasks require product functions and functions satisfy needs. Needs provide the channel to VOC.

You cannot proceed forward until you know what level of VOC data you currently have. Assemble the current VOC data and search for gaps to fill and uncertainties to verify.

Next identify the data sources. Some of these were listed earlier in this chapter. Final selection of data sources will depend on urgency, budget, and time to do as mentioned earlier.

Plan the collection. Selection of the method can be important. How many customers do you talk to? Who does the collecting? Do you use your own people or hire from the outside? This is one of the most time-consuming parts of the VOC process. Good planning here will pay off later.

Once the data are in, it must be organized and structured. One useful method to structure VOC data is to categorize it using affinity diagrams discussed in Chapter 1. Tree diagrams are also quite useful because they can be used to enter the data directly onto a QFD matrix. An absolute must at this point is to draft a glossary of terms in order that everyone uses the same definitions and speaks the same language.

Until now we have developed a qualitative VOC with a set of structured, ordered, standardized customer verbatim labels. Now we must bring magnitude to this order by quantifying the verbatims for importance. Several value measurement techniques are appropriate for this task such as category scaling, the analytical hierarchy process discussed in Chapter 4, or the constant sum method of allocating 100 points across verbatims. Who does the importance rating? Too often it is done internally by the team. Be careful of this method. How can the team quantify what they were originally intending to get from the customer? The customer is really the one to rate the items. Too often we don't take the time to go back to the customer interviewees. This almost always leads to trouble. You can rate the verbatims internally on your own but sooner or later the importances should be verified with the customer.

The VOC items are now ready to be translated into product technical parameters. This translation converts the VOC into a language and terms that are more meaningful to engineers and technical people. Two translation mechanisms are the HOQ and COPC.

The VOC process is open-ended and consists of feedback loops to keep the data current through customer verification and correction. Figure 9.13 shows the feedback loop exiting at the quantification function and/or the translation, QFD, function. Quantification and translation always raise issues. There can be gaps in the data, confusion in the glossary, and confusion over the importance ratings. To clarify these concerns it is necessary to go back to the customer. Since the first data collection, new data sources may have come into being or the business plan may have changed. The feedback loop can be reentered at the data source functions or collection function or both.

SUMMARY

In this chapter I have shown how the VOC fits into the scheme of supply and demand and the fact that there is some order to all of this chaos. VOC consists of both qualitative and quantitative data. Part of the order of things is illustrated by the fact that it is necessary to identify the labels on customer verbatims before we can use the quantitative methods. Too often teams want to reverse this order. Reverse order is also too often used in collecting VOC data. That is, teams want to collect VOC before doing the HOQ or a COPC matrix when they don't know what questions to ask. Do the HOQ first with existing data, then design and ask VOC questions. Defining the customer is a considerable challenge sometimes. Too often we want to build a product that is all things to all people. You can't do that anymore and still stay in business. I describe a customer morphology matrix to organize customer information to build a boundary around the target operating arena. This boundary serves as a guide to define the customer in order to define VOC. Other useful tools are GOAL/QPC's VOCT and Arbor's Customer Window℠. They provide another perspective for documenting or expanding VOC. I emphasize the importance of listening carefully to what the customer is saying. We have to be alert to things like solutions to problems masked as customer needs. After the VOC labels have been established, it is important to build a glossary of terms to aid communication and increase consistency of interpretation among team members.

VOC and customer needs change with time. Periodic checks should be conducted to gauge the trends in importance in customer needs. The Delphi questioning method is a very useful tool to elicit trends and likelihood of occurrence. Monitoring is an effective surveillance method for tracking VOC verbatims and events into the future. It is also discussed in Chapter 6.

In QFD and COPC, very little moves without the VOC. A good job on VOC early on in the QFD time frame can be well worth the extra time and patience it takes to do.

The most important point to remember is that the VOC is not just the voice of the customer, it is not just the customer verbatims or the stacks of survey sheets, but it is the integrated and processed understanding of all of these data. The collected data alone cannot do the job. They must be interpreted and woven into the development cycle through a structured process. QFD and

COPC are the translation mechanisms to weave the VOC into the fabric of the company commercialization process.

VOC and QFD or COPC are two different processes. VOC can be collected and structured without having anything to do with QFD. VOC can stand alone. QFD/COPC on the other hand cannot operate without the VOC. VOC is an independent process whereas QFD and COPC are a dependent process. They depend on VOC. Remember, QFD is a structuring tool it is not a collection tool!

REFERENCES

1. Shillito, M. L., and DeMarle, D. J., *Value: Its Measurement, Design, and Management,* John Wiley & Sons, New York, 1992.
2. Marsh, S., Moran, J. W., Nakui, S., and Hoffherr, G., *Facilitating and Training in Quality Function Deployment,* GOAL/QPC, Methuen, Mass., 1991.
3. Cary, M., Kay, B., Orleman, P., Robertson, W., Ross, G., Saunders, D., Wallace, W., and Wittenbraker, J., "The Customer Window," *Quality Progress,* pp. 37–42, June 1987.
4. King, R., *Better Designs in Half the Time, Implementing QFD Quality Function Deployment in America,* GOAL/QPC, Methuen, Mass., 1987.
5. Jolson, M. A., and Rossow, G. L., "The Delphi Process in Marketing Decision Making," *Journal of Marketing Research* **8,** 443–448, November 1971.
6. Salancik, J. R., Wenger, W., and Helfer, E., "The Construction of Delphi Event Statements," *Technological Forecasting and Social Change,* **3,** 65–73, 1971.

BIBLIOGRAPHY

1. Adams, R. M., and Gavoor, M. D., "Quality Function Deployment: Its Promise and Reality," in *Proceedings, 5th GOAL/QPC Conference,* GOAL/QPC, Methuen, Mass., November 1989.
2. Estes, G. M., and Kuespert, D., "Delphi in Industrial Forecasting," *Chemical and Engineering News,* pp. 40–47, August 23, 1976.
3. Griffen, A., and Hauser, J. R., "The Voice of the Customer," *M. I. T. Marketing Center Working Paper 91–2,* Sloan School of Management, M. I. T., Cambridge, Mass., January 1991.
4. Hertig, J. C., and Abbott, M., "QFD in Pharmaceuticals, A Case Study," in *Proceedings, GOAL/QPC 6th Annual Conference,* Boston, Mass., 1989.
5. Klein, R. L., "New Techniques for Listening to the Voice of the Customer," in *Proceedings, 2nd Symposium on Quality Function Deployment,* Novi, Mich., June 1990, pp. 197–203.
6. Nakui, S., Marsh, S., and Ono, M., "Hearing the Voice of the Customer," *Competitive Times,* GOAL/QPC, Methuen, Mass., 1991, pp. 5–10.
7. Vanston, J. H., Jr., *Technology Forecasting: An Aid to Effective Technology Management,* Technology Futures, Inc., Austin, Tex., 1982.

CHAPTER 10

SOME PARTING COMMENTS

HUMAN DYNAMICS AND CROSS-FUNCTIONAL INVOLVEMENT

QFD is a problem solving and a planning process. As such it is a dynamic process requiring review and revision. The process is iterative and loops back as needed, refining and upgrading the results as it pulses forward. QFD requires interdisciplinary participation and hierarchical involvement. QFD can become complex because it cuts across organizational lines and functions. However, its structured approach helps break complexity into smaller manageable pieces so they can be integrated into a better design.

MISCONCEPTIONS

The QFD process is so deceptively simple, reasonably straightforward, and intuitively obvious that it can be labeled as motherhood or many people can respond that they "are already doing it." As such, users may begin to take the process lightly. The misconception is derived from the fact that people have QFD-type data that look similar to that in QFD projects, but they do not have it in a QFD matrix to structure it and reveal relationships so it can be interpreted. It takes the matrix process to turn data into information. At the end of completing a QFD exercise comments also appear such as, "I could have drawn the same conclusions months earlier." I do not agree with those comments but I can understand why they are sometimes made.

The theoretical QFD model is simple. It is the translation into reality that makes the process complex and often abandoned, leaving long lasting negative impressions about QFD with those involved. The individual pieces are relatively easy to establish, but the integration of the pieces up and down the organizational hierarchy and across functional lines, into a consistent compatible set of plans, is the cause of frustration frequently encountered in the process.

SHORTCUTTING THE PROCESS (QUICK SOLUTIONS)

Here's where problems begin. Many managers commission a QFD project thinking they are going to get quick results. They, many times, apply QFD too late in the product life cycle when they are in a panic to get the product out. In such situations the first tendency is to shortcut the QFD process. By doing so things don't work out as anticipated and QFD and the process facilitator get a black eye. There are times when it is best not to do QFD if it is applied too late or if it is used for quick solutions. I admit that the QFD process takes too long. I and many others are trying to develop a shorter method without sacrifice to the overall process. This has been a difficult task and I have not yet found a solution. So far the COPC process seems to be the best compromise. Even COPC is not that much shorter. However, because it gets to technology and design more quickly, it is perceived as being shorter and people have less anxiety about taking it on.

A faster way to do a HOQ is to work at the next level up in the QFD tree diagram (see Chapters 1 and 4). That is, instead of working at the level 3 verbatims that may consist of 200 to 600 customer needs, the team would work at the next upper level which may contain 30 to 50 needs. This level of needs may also be reduced further by using only those needs that have a customer importance rating of "4" or "5" (based on a 5 point scale). The same procedure could be used for the HOQ columns. Doing so, the team will be working with the Pareto vital few. Working with a detailed matrix like 100 by 100 is enough to give the team members a heart attack!

INTEGRATION—KEY TO THE FUTURE

The key to QFD's future is integration. I have tried to show examples of this in Chapters 4, 5, and 6, where I discuss VE, planning, and TF. QFD will also be embedded in company total quality management, reengineering, and rapid commercialization programs as well as many other quality programs. In this respect there are two important types of integration. First there is integration of technologies to improve the model or the process. Then there is integration within other programs and philosophies like TQM, concurrent engineering, and so on.

QFD used alone in its purest sense will not survive the decade. It will pass as if it were a "program of the month." Practitioners will have to be creative and flexible in its use.

COMMUNICATIONS

QFD, in addition to all of its other virtues, is a communications process. In order for people and groups to communicate we need two things. We need a structure and we need a language. The various QFD matrices provide the structure; the numbers that we put into the matrices provide us with a numerical language. For example, what one team member rates a 9, another member may rate a 2. At this point we have the groundwork for dialogue. Those members who rated an item a 2 share with the rest of the group the reasons behind their rating. Likewise, those members who rated it a 9 also share their reasons. After feedback and discussion, the team rates the item again to see if there is convergence approaching consensus. QFD makes opinions measurable. The dialogue required to resolve such disparities means more than the final resulting numbers. This dialogue is invaluable and probably would never take place if the QFD process is not used.

Consensus need not be mandatory, as discussed earlier in Chapter 1. Getting consensus on every piece of the HOQ and the other three matrices would take forever. This is a direct contradiction to what is published in most literature. I have found it just as effective to go with the majority and write a minority report or footnote stating the reasons for the disagreement. This way both sides of the argument are preserved and documented and can be resurfaced at any time.

WHEN TO DO QFD/COPC?

By now you are wondering, which is the correct way to do QFD or COPC? After all, you probably read many articles and talked to many practitioners only to find out that many people apply the processes differently. For example, some teams will not begin a QFD until they do focus groups and surveys. Others, like me, do the HOQ first and use the missing data to design focus groups and surveys, as discussed in the next section. There is no single correct way to apply all of the tools. The correct way is the way that works best for the team for the particular application. The worst situation is to force the application to fit a rigid process lead by an inflexible process facilitator. Consider the tools discussed in this book as the basics and adapt from there.

Starting a QFD other than in the very beginning of the commercialization process or the product life cycle will be an exercise in frustration. The team will not back up to start over to do a QFD, and the QFD facilitator has only pieces to build the HOQ and other matrices. There are exceptions, but, generally this is a lose–lose situation. I have personally helped teams do a reversed COPC by taking a product concept and checking it with customer and company criteria; however, I will do my best to make this an exception rather than a rule.

QFD, A STRUCTURING TOOL, NOT A COLLECTING TOOL

Too many people have a misunderstanding that QFD is a proactive customer needs collecting tool. QFD is, instead, a structuring tool that allows already collected data to be structured and quantified, in order to show relationships between data. The signals from the hierarchical data are then used to determine what data to verify and what new data to collect. Figure 10.1 illustrates the process. Data are developed in a random fashion. They are then structured using affinity diagrams and matrices. Next, order and hierarchy are developed through the use of rating scales. The hierarchy puts us in a position to converge on specific issues in order to bring focus to planning efforts. The output of the process is then given to marketing/business research to develop a focused inquiry to verify existing data, answer questions about soft data, as well as to develop new data we didn't know we were missing.

Too often clients think they have to do surveys and other market research before doing QFD. I generally discourage this because, how do they know what questions to ask? Do the rows of the HOQ first, based on current available data to determine what you know and what you don't know. Then, based on the current state QFD, determine what questions to ask and from whom.

HIGH POWERED MATH AND STATISTICS TO THE RESCUE! BEWARE!

One of the more dangerous operations performed with the QFD data and matrices is to apply higher order statistics to the matrix of numbers. This involves the various cuts and sorts, correlations and regressions of the type usually reserved for business research. First of all, QFD is NOT business research, nor is it a marketing or marketing research tool. If one uses the higher order math, one does so at one's own risk. Matrixes full of numbers are black holes for statisticians to get sucked into. The majority of the data in the matrices are highly subjective in nature and can lead one to spurious results and false positives. Therefore, if you insist on using these mathematical tools with QFD data, be careful! It is too easy to make mistakes. It is easy to go on a witch hunt!

Remember, QFD is an communication process. The matrices and numbers allow us to develop good debate. Statistics can do this also, just as long as the team and other users of QFD statistics realize that the resulting mathematical manipulations are first just that, a dialoguing technique. They should be used in the same capacity as the tools discussed in Chapters 3, 5, and 6.

ACCURACY AND FINE TUNING

How accurate should a QFD or COPC be? How precisely should we refine the numbers in the matrices? What is accuracy? Remember, QFD and COPC are

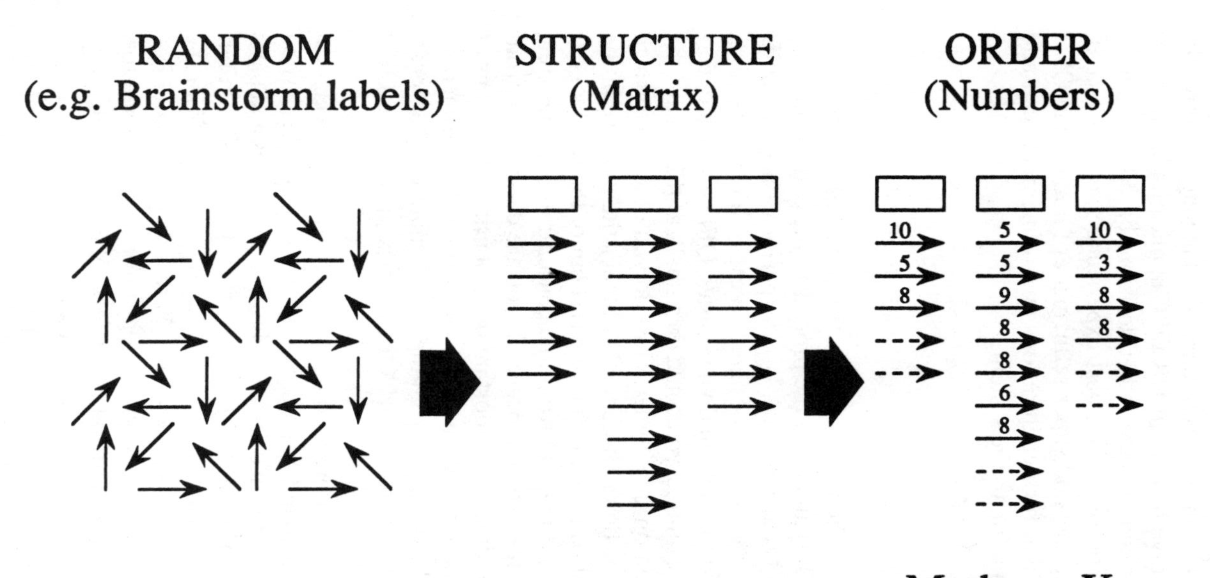

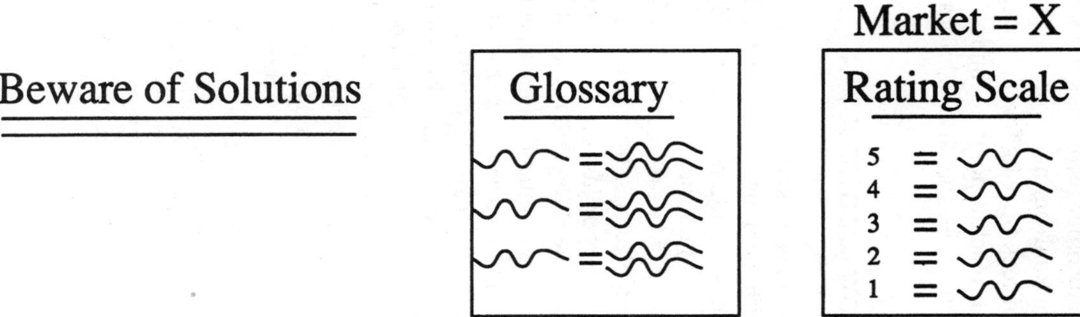

Figure 10.1 The Structure of data

dialoguing techniques. Although the matrices contain many numbers, the process is not a true mathematical technique. The process is a numerical technique used in a qualitative capacity. As mentioned earlier, QFD and COPC are structuring techniques that use numbers as a numerical language. The resulting recommendations that evolve from a QFD/COPC application are a set of many other recommendations coming from other decisions based on the integration from numerous decision processes, past experience and, let's face it, gut feel. Striving for high accuracy and super refinement of numbers in QFD can reach a point beyond which further iterations and refinements can add more noise than useful information to the system.

HOW MANY QFDS?

We have talked at length about the QFD process. We discussed how to begin with the house of quality (HOQ) and drive the process to the end. At no time did we mention how many QFDs to construct. Do you perform one large QFD (HOQ) on the entire product? If you are designing a car or a copy machine, this could be a gigantic matrix! Or, do you do a separate QFD on each of certain selected subsystems? For example, you could do a smaller QFD on just the car door or just the toning station in a copier. How do you determine which and how many QFDs should be done? It depends whether the purpose of the QFD is product design or problem solving (getting rust out of the car door). If the purpose is problem solving, the QFD would be concentrated on the area, function, or subsystem where the problem occurs. If you are designing a product, you would concentrate on the most important or most critical functions or subsystems for product operation and customer and company satisfaction. This information can readily be obtained from the COPC matrix. After locating the most important functions from the COPC matrix, a separate QFD would be performed on each one.

A HOQ would be constructed in order to determine the product technical requirements (PTRs), as discussed in Chapters 1 and 2. PTRs would be carried forward through the HOQ relationship matrix to Matrix 2, the pieces and parts matrix (Chapter 2). In Matrix 2 (Figure 10.2) the important functions and their related technologies for a particular path would be carried forward from the COPC matrix and entered as columns in Matrix 2. Parts and pieces would be derived (brainstormed) from the function-technology information carried forward from the COPC matrix. This usually takes the form of a team brain dump to fill each function technology with parts and pieces. However, there is lots of structure to do so. Normally, in regular QFD, the parts and pieces are brainstormed against a concept from a Pugh concept selection matrix. This can be difficult because the concept has not been divided into functions at a subsystem level. This matrix is now equivalent to the original Matrix 2 in the traditional QFD model discussed in Chapters 1 and 2. The team still uses this matrix to proceed forward to the other two QFD matrices.

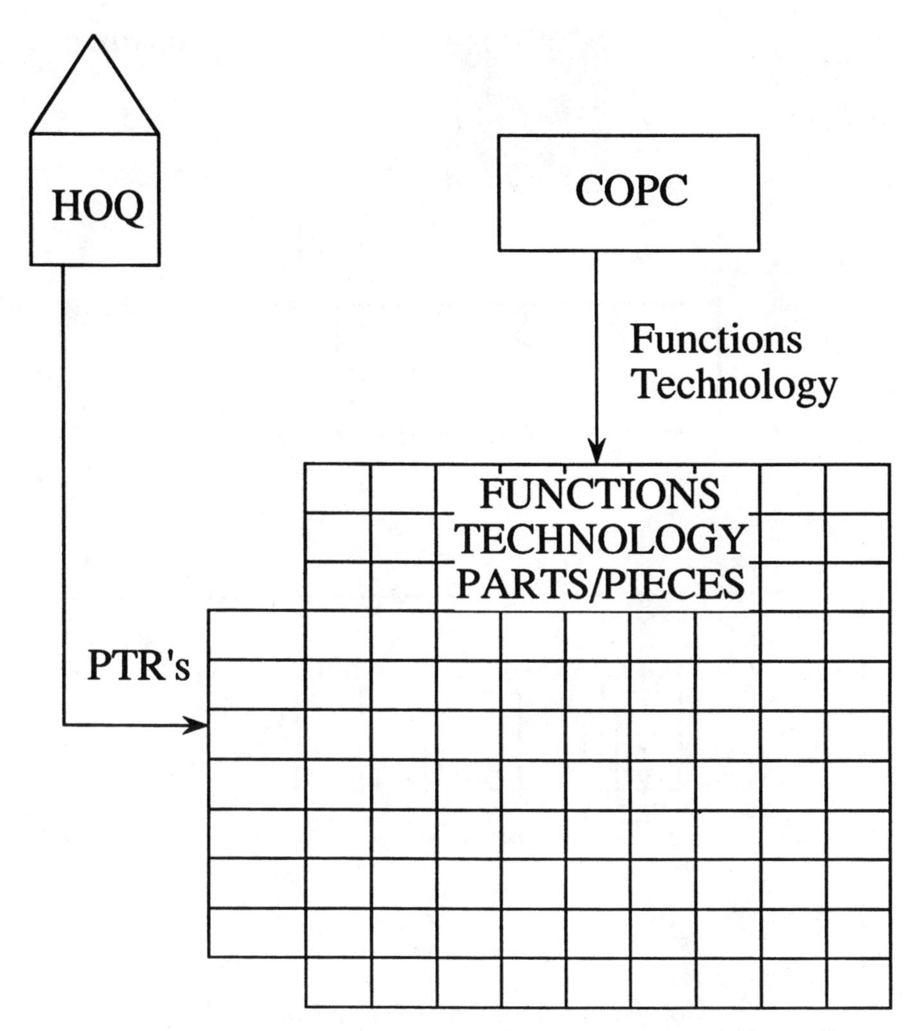

Figure 10.2 COPC entry to the QFD process.

Figure 10.3 illustrates the connection between COPC and the QFD process. The three layers, strategic, tactical, and operational, must all be connected and in sync from top to bottom and vice versa. So often QFD projects are started with little regard for the upper two layers. Too often exemplary QFD projects have been conducted on the wrong product, subsystem, and so on. The process begins by answering the 12 questions discussed in Chapters 4, 7, and 8. This forces strategic focus, which becomes an overlay for the COPC at the tactical level. The concepts that surface from the COPC process are now translated into the company commercialization process through multiple small QFDs (HOQs), as opposed to one huge HOQ. These small QFDs may also be based on the most critical functions in the COPC matrix. For example, instead of

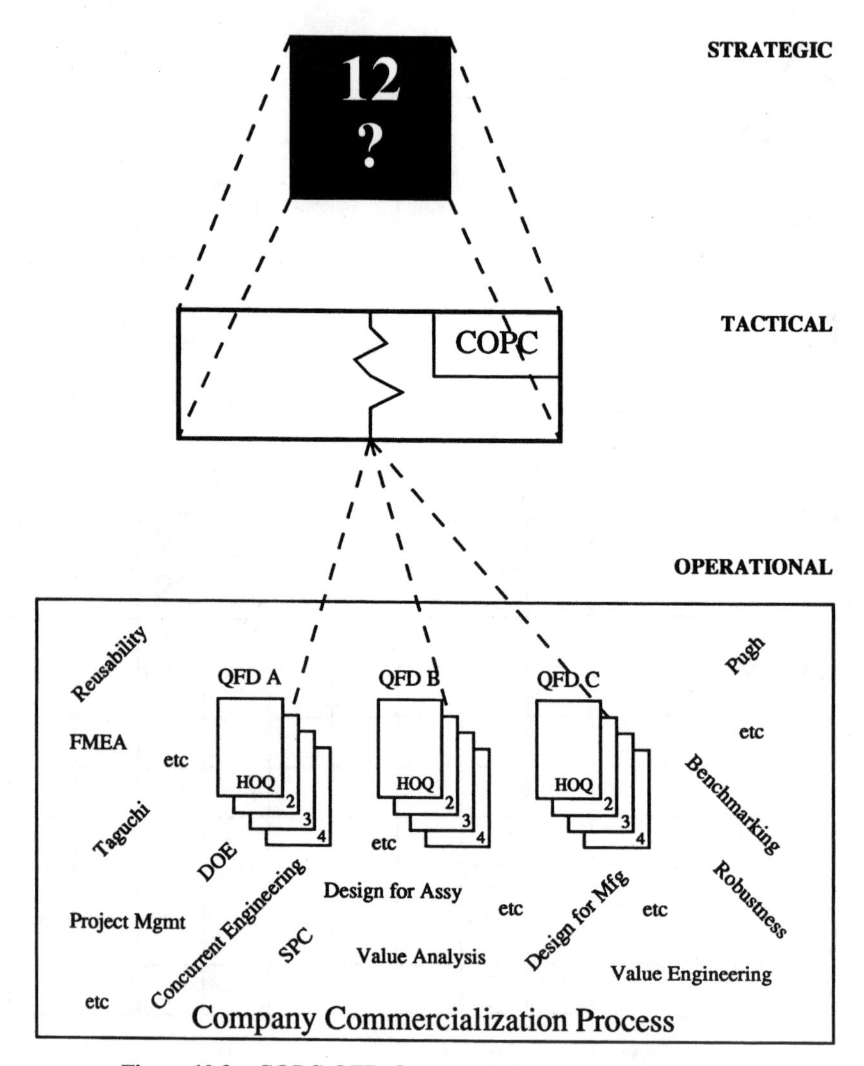

Figure 10.3 COPC-QFD Commercialization Process Interface

doing one large QFD on a copier, perhaps two smaller QFDs are constructed, one on the toning station and another on the paper feeder.

Most of the literature on QFD mentions little or nothing about connecting the QFD matrices to the company business plan, market participation strategy, mission, and so on. Chapter 5, on PQFD, discusses a method for connecting all of the layers. Too often companies do not have a good bridge between these layers. COPC generally provides that transition in the absence of a formal process.

Notice also that QFD is illustrated as one of many tools used within the company commercialization process.

MANAGEMENT PARTICIPATION

In order for programs and processes like TQM and QFD to work in a company, management must participate—not just support and not just be involved, but participate! Management's verbal support is just not enough; it doesn't work. Programs that have only verbal support evaporate quickly.

Management must get involved through participation. Unfortunately, there is great reluctance to do this. The old excuse, "we are busy and don't have the time" just does not fly anymore. To stay in business, management is going to have to rearrange priorities to allow participation in QFD- and COPC-type activity.

I believe one of the reasons management does not get more involved in QFD processes is that they don't know how to become involved and, in some cases, don't understand the process itself.

If a management participation process is in place to encourage and enable managers to comfortably participate in the process such as attending QFD/COPC team meetings, I believe we would see more management intervention. Such a process would have a participation kit available to all managers. This kit would contain a series of questions that a manager could ask when attending team meetings. In fact, a wallet card could be developed that would contain the intervention questions. This would allow managers to intervene before crucial decisions are made. They would make course corrections before it is too late or before enormous amounts of money have been expended. A short (less than one hour) seminar could be developed for management explaining how they can participate in QFD activity. The above kit would be part of the seminar.

When participating in a COPC/QFD team meeting, a manager would ask to see the following:

1. Market/customer document (who is the customer?).
2. Manufacturing decision criteria (what are they and how important is each?).
3. Voice of the customer document (customer needs and importance).
4. COPC matrix.
5. HOQ matrix.
6. Scoring model process for rating technology.
7. Customer matrix.

The following questions could be used to promote dialogue and check the direction of the team:

1. Have you checked your product concept proposals with the business plan? Does it fit? Why? Why not? Do you have a copy of the business plan?

2. How did you choose country, market, segment, and user? Did you check this with management? Marketing? Did marketing participate? Why? Why not? Does this fit the company mission and business? Why? Why not?
3. How did you arrive at the importance of the VOC items? Did you use an interdisciplinary team? Was marketing involved? Did marketing participate? Who? Why? Why not?
4. Are the elements of the business plan consistent with the corporate mission?
5. What can you do to elevate some of the weak scoring elements of the product concept to a better score? What will it take technologically? How much would it cost? How long will it take? Is an invention required? Can we reuse existing technology?
6. How does your proposed concept compare to our competitor? Can we win? Will they win? Can we catch up? Can we leap beyond? Do we need more detailed benchmarking? What is the reason for the gap between us and them? Can it be closed? What will it cost? How long will it take?
7. How does your proposed concept and resulting technology support the technology platform? Can the technology be leveraged into the future?
8. Is the importance of the VOC and customer verbatims static? Will they change with time? Has this been considered? What will be the importance three, six, and ten years from now? Can this be reflected in the current choice of technology?
9. Have you used industry learning curves to estimate unit manufacturing cost and street price at time of introduction? Have you used industry experience curves to estimate competitor's cost and price at time of our introduction?
10. What data led you to select your particular alternative concepts/recommendations?
11. What is the quality and/or performance improvement potential for the product as a result of these changes/recommendations?
12. Have you found any problems in the PTR correlations that hinder progress or cause problems?
13. Are there any inconsistencies between the VOC and any of the target values?
14. What is your confidence around the VOC importance ratings? Which ones are you not sure about? Will these be verified? How?
15. Where is your current product weak/strong compared to the current competitor's product?

Xerox has addressed management participation very nicely through their LUTI model.[1] The LUTI (learn-use-train-inspect) model is a management tool for cascading training from upper management to shop floor. First, the first level of management learns, for example, COPC by taking a course. They then use the process on one of their projects. Next, the managers train their direct reports in the methodology. This may require help from the in-house process experts. Finally, they inspect for appropriate application of the process during their regular operations or staff meetings. The next level of management then executes LUTI and the training methodology starts again until it quickly cascades down the organization.

Having gone through the LUTI process, management is no longer in the dark with regard to process: there should be no reluctance to participate in team meetings. They will know process and they will know what to ask.

Whether you use the LUTI approach or a similar model, management needs a user-friendly mechanism to become involved in QFD and commercialization processes. Without this participation any such process or program will be on shaky ground.

WHAT SIZE MATRIX?

How large should a QFD matrix be? If it is too large participants will drown in data and starve for information. Yes, I have seen a 400 × 1000 house of quality! It makes one cringe just to think about it. The QFD project was completed and was successful, but there was a lot of pain and suffering along the way. Some of the QFD team members will never volunteer for or participate on a QFD team again. We lost them. We cannot afford to have these kind of stories floating around.

The mistake the team made was to cook the elephant all in one piece. A better approach is to break the product into smaller pieces and cook the elephant one piece at a time. Which part do you work on first? How do you break the product into smaller pieces? Which pieces are more important than the others? How do we measure the importance of the pieces?

One way to break the product into smaller pieces is to work from the COPC matrix as discussed earlier under the section "How Many QFDs?" A QFD matrix would be constructed on just the most critical functions in the COPC matrix. These matrices should be no larger than 20 × 20.

Another method is to break the product into its corresponding subsystems. For example, a copier would have about 18 subsystems. Value measurement techniques can then be used to quantify the importance of each subsystem. One measurement method to use is the 100-point sum-to-unity rule. That is, the team allocates points across the subsystems such that they add to 100. This method was discussed earlier in Chapter 4 under COPC, Step 10. It forces trade-offs and readily surfaces the more important items.

Another value measurement method is Pareto voting,[2] wherein participants are allowed to vote for less than 20% of the total subsystem. In our copier above, I would allow the participants to select the three (18 × 0.2 = 3.6) most important subsystems. Only one vote is allowed per subsystem. People may not apply all votes to a single item. The total votes for the group are then tallied to form a Pareto distribution. There are numerous value measurement and value screening techniques that can be used and are discussed at length in Reference 3, Chapter 4.

A separate QFD matrix would be constructed for the vital few Pareto subsystems, the most important ones. A rule of thumb for matrix size is 20×20, as mentioned earlier. This may entail going to the next higher level in the tree diagram (discussed in Chapter 1). Try to avoid making gigantic matrices that collapse from information overload. Remember, you can't cook an elephant all in one piece!

PARADIGMS ABOUT PROJECTS

In order for the commercialization system to work, that is, the team and the process, we are going to have to change our thinking and shift some paradigms. We will have to take a hard look at how the new product commercialization process and the associated best practices work in the company. To compete, internal business may never be the same. Over the years I have experienced and collected some classic paradigms about projects. They seem to appear over and over again. All of these paradigms generally lead backwards and drag out the commercialization process. I have listed these paradigms below along with the future counter-paradigm. Some may refer to this comparison as a culture change, maybe so. To me they are the elements of survival.

PARADIGMS ABOUT PROJECTS

Old Paradigm	New Paradigm
Just get the product out the door, we'll fix the problem later.	We cannot ship a product before it is ready. Schedules may have to be adjusted to accommodate this.
We just keep making changes and additions right up to when we are ready to cut metal—that way we will have incorporated everything in product.	Any additions or changes after a certain date will be put into Model #2.
We have to hit a home run with every new product; therefore, we will wait until we incorporate all the latest high-tech stuff.	We can't hit a home run on every product; therefore, we will have to make several base hits and come out with small updated models every 6 months. That way, we can get to market faster and not have to catch up.

We cannot afford to use processes like COPC, QFD, Value Engineering because they might push us off schedule and, besides, we need quick results.	Check with management about firmness of schedule and ship date—they may agree to ship schedule since these processes will save us money and cut the remake frequency.
We need quick results; therefore, we must short-cut the QFD process.	We can no longer afford to leap before we look. We must use a structured process to help us become effective before we are effiicient.
It is important to quickly come up with a design, then we will look for a customer because anything we have our logo on will sell!	"Ready, fire, aim!" is a sure path to organized oblivion. Companies have a very good track record at firing first. This usually results from having no target. The logo is no longer a guaranteed sell.
We will try to do this project ourselves otherwise business XYZ will get the sales credit.	We must all work together otherwise we may not be able to develop the right product because we may be too narrowly focused.
We know what the customer wants better than the customer!	This is arrogant suicide!
We can't tell the decision maker our real recommendations because they are different from the bosses original plans.	The wages of "brown nosing" are slow suffocation. We must face the facts.
We need the product model to show at the trade show, we'll make it work later.	If it doesn't work at the trade show, we may not have to make it work later.
I am using this project as a stepping stone for a promotion. I need to make decisions that look good and QFD may reveal the truth.	You can pay me now or pay me later.

To practice the new paradigms will require everyone's help, cooperation, and courage. Process facilitators, in particular, QFD facilitators, are in a particularly good position to draw attention to the old paradigms when they surface. A neutral third party can say and do things that many team members cannot do. Paradigms can be subtle and insidious. Many companies will require a major culture change to use all of these tools like QFD to survive. One place to start working on that change will be with project teams using the many tools and practices discussed in this book. Let's get started!

SUMMARY

QFD is a way of looking into the future with respect to products and markets. It is a planning tool based on structure, order, and prioritization. But, most of all, it is a communications tool, a vehicle for dialogue. It is a long term oriented process that does not produce short term solutions. It cannot be applied at the last minute with preconceived intentions of shortcutting the process. QFD works. It works better when supplemented with other techniques like value ratios and value graphs. Technological forecasting tools can be used to locate and test the consequences of QFD signals. QFD's logic of interlocking matrices can also be used in a planning capacity (PQFD) to develop a better connection to corporate planning. Because QFD incorporates the use of interdisciplinary teams, the behavioral and organizational aspects are as important as the QFD process itself. Too often QFD users do not take advantage of all that can be done with a completed QFD or COPC. To succeed, our applications of QFD will have to be creative. We will need a paradigm shift in application and process design. We will need principled disobedience to established paradigms rather than docile obedience. Unfortunately, it is a shame that most companies have to degenerate to a panic survival mode before they will use processes like QFD. Pressure seems to be the only thing that motivates.

Remember, no one of the tools discussed herein can be used alone. The practitioner should pick the combination of tools and enhancements that work best for a particular situation. The same precise combination of tools will rarely be used repeatedly.

The future will provide more than enough opportunity for QFD to last a lifetime. The opportunity will come not as distinct signals for pure QFD but will come as coded messages requiring integration of QFD with all of the tools and techniques to bring about a successful future. The choice to integrate with the future is entirely yours. You can choose to change your paradigm.

REFERENCES

1. Watson, G. H., "From Theory to Application: Bridging the Quality Implementation Gap," *Proceedings, Rochester Section ASQC, 49th Annual QC Conference,* March 1993, pp. 285–297.
2. Shillito, M. L., "Pareto Voting," *Proceedings, Society of American Value Engineers* **8,** 131–135, 1973.
3. Shillito, M. L., and DeMarle, D. J., *Value: Its Measurement, Design and Management,* John Wiley & Sons, New York, 1992.

INDEX